AF788738

Instructor's Guide

Microscale Experiments for General Chemistry

Kenneth L. Williamson
Mount Holyoke College

John G. Little
Saint Mary's High School
Stockton, CA

HOUGHTON MIFFLIN COMPANY BOSTON NEW YORK

Senior Sponsoring Editor: Richard Stratton
Assistant Editor: Marianne Stepanian
Manufacturing Coordinator: Marie Barnes
Associate Director of Marketing: Karen Natale

Printed in the U.S.A.

ISBN: 0-669-41607-X

2 3 4 5 6 7 8 9-CRS-03

Table of Contents

Using This Guide: Some Introductory Notes

Intent: In preparing this guide, we have tried to meet the needs of a diverse group of users: graduate student teachers' assistants (TAs) responsible for the setup for their own sections, laboratory managers responsible for preparing reagents and for setting out special apparatus, and laboratory instructors who must oversee the actual carrying out of the procedures.

Preparation Tips: Reagents are listed separately for each experiment, and we have indicated the quantities needed per section (based on 24 students, working individually). Directions have been given for preparing solutions, even for those reagents, such as 3 *M* HCl, that are often available without special preparation. The directions will almost always give quantities greater than what will be needed for a single section both because we recognize that most schools run many sections and because it is more convenient to prepare more than one expects to need, especially for small volumes of dilute ($\leq$0.1 *M*).

We have provided estimates, based on our own experience in using these experiments in our own classes, of the time needed to set up each activity and the time the average student will need to complete it. In doing so, we have assumed that the person(s) doing the setup have a good background in the theory and practice of typical chemical experiments, are only mildly familiar with the special considerations involved with micro- and small-scale activities, and are generally unfamiliar with the specific activities in this book.

Although we have tried to be as complete as possible in directions to be followed by the students, in the interest of clarity, we have occasionally left some of the instruction to the laboratory instructor. In such cases, we have included the special instructions in the section entitled "Notes on the Procedure."

Safety Notes: Both of us are very safety conscious, although neither has had any major mishaps, for which we are thankful. One (JGL) is almost fanatical on the use of eyewear that provides full splash protection at all times in the laboratory, regardless of the wearer's reason for being in the laboratory.

In giving directions for solution preparation, especially for diluting acids and bases, we have chosen to urge extreme safety precautions; better to err on the side of safety than to risk debilitating injury.

Answers to Questions: All Prelaboratory and Postlaboratory Questions are answered in this guide.

Sample Data: Sample data are provided for all quantitative experiments, although it is reasonable to expect some variation. In those cases where data are particularly sensitive to technique, or where results consistently vary from expected values, we have noted this fact. Further, we have made an effort to adhere to the standard rules for significant figures.

EXPERIMENT

1 Observing a Chemical Reaction

Placement: First experiment of the year

Approximate time for completion: 30 minutes

Approximate preparation time: 35 minutes

Materials: 24 of each unless noted otherwise, assuming a class of 24 students working individually

Apparatus

Safety goggles
96-well test plate
1-mL graduated pipette
Glass rod or plastic toothpick
Thermometer, –10 to 110°C
Reaction tubes or 10 x 75 mm culture tubes, three
Beaker, 100 mL
Ring stand
Small three-prong clamp. If a standard utility clamp is substituted, it may be necessary to wrap the tube for Step 2 in paper to provide a snug fit.
10-mL graduated cylinder
50-mL beaker
Magnifying glass (optional; can be shared)
10-mL Erlenmeyer flask
Wax pencil (3 or 4) (needed only if ungraduated tubes are used in Step 2)
1-mL graduated Beral-type pipettes, one for each stock container of 1 *M* $CaCl_2$, 3 *M* H_2SO_4, and 1 *M* Na_2CO_3 solutions; recommended for 1 class of 24, 2 sets of stock bottles, 6 pipettes

Reagents

3 *M* hydrochloric acid, HCl, 10 mL. Slowly and with constant stirring, add 25 mL of concentrated (12 *M*) acid to 75 mL of distilled water. Watch for signs of overheating. *Wear gloves and face shield. Work in the fume hood.*

3 *M* acetic acid, $HC_2H_3O_2$, 15 mL. Slowly and with constant stirring, add 18 mL of glacial acetic acid to 82 mL of distilled water. While acetic acid is not as strong as hydrochloric acid, precautions similar to those used with hydrochloric acid should be taken.

Steel wool, #0000, 15 g. Steel wool (iron) is sold with a coating of oil to prevent rusting. This must be washed off in ligroin, hexane, or another nonpolar solvent shortly before use. Be sure the washed steel wool is dry.

1.5 *M* sodium bicarbonate solution, $NaHCO_3$, 25 mL. Dissolve 8.4 g in 100 ml of distilled water.

1.0 *M* calcium chloride solution, $CaCl_2$, 25 mL. Dissolve 11.1 g of the anhydrous solid or 14.7 g of the dihydrate ($CaCl_2{\cdot}2H_2O$) in 100 mL of distilled water.

3 *M* sulfuric acid, H_2SO_4, 25 mL. Slowly and with constant stirring, add 17 mL of concentrated acid to 83 mL of distilled water. *Caution: Very exothermic; wear gloves and face shield, and work in the hood.*

1.0 *M* sodium carbonate solution, Na_2CO_3, 100 mL. Dissolve 10.6 g of the anhydrous solid of 12.4 g of the monohydrate ($Na_2CO_3{\cdot}H_2O$) in 100 mL of distilled water.

Discussion

Students are introduced to and asked to interpret common phenomena (many of which are familiar reactions in unfamiliar context). No pretense is made that all evidences of reaction will be met here; rather, the emphasis is on laying a foundation of recordkeeping, distinguishing between qualitative and quantitative observations, and introducing basic techniques of data analysis.

It is left to the instructor to provide any needed definitions of terms such as *deviation* and *mean*. The same data in Prelaboratory Question 1 may be used for instruction purposes if desired.

Notes on the Procedure

Step 1. Some effervescence should be evident with the HCl, but not with the acetic acid, notwithstanding equal concentrations. On the other hand, only the latter will have any appreciable odor, so there must be some sort of vapor present in both. Over time, the steel wool will be discolored and partly decomposed by hydrochloric acid, but the acetic acid will leave the metal undisturbed. Students might be interested to know that the acetic acid corresponds to roughly triple-strength vinegar.

Step 2. As noted above under "Materials," the steel wool must be rinsed free of its protective oil coating and then dried before use. Except under extreme conditions of humidity, this can be done several hours or up to a day in advance without risk of premature rusting. The vinegar provides a mildly acidic, humid environment in which the iron will rust, absorbing O_2 from the air. Removal of oxygen from the atmosphere in the tube causes the water level to rise, reaching a maximum in about 30 minutes. Usual results are a bit short of the actual percentage of O_2—results run on the order of a 15% decrease in volume of trapped air.

If you have them, the calibrated reaction tubes are a definite advantage for this part of the procedure. If unmarked tubes and wax pencils are to be employed, use the narrowest diameter available to maximize the change in water column height. Tell students to briefly insert the tube with the steel wool plug into the beaker of water, but without clamping it in place. They then note the level and then quickly remove the tube, dry the mouth, mark the level, and reinsert the tube in the beaker, this time clamping it in place.

Steps 3–5. This is an introduction to conservation of mass and to the fact that it only applies to a closed system. Two chemical changes are observed, but mass is conserved only for one (precipitation), since the other involves release of a gas (CO_2). Some students will be surprised to find that bases have mass and may ascribe the observed loss of mass solely to the spraying they observe.

Step 6. As noted above, expect about a 15% decrease in volume. Reaction tubes permit direct volume readings, so students have little trouble. With plain tubes, students mark the new height of the water in the tube and then determine the percentage decrease in air volume either using the assumption that column height is directly proportional to volume or by actually measuring the volume of water needed to fill the tube to each mark. (The latter method does not give significantly better results but is a good exercise, especially if the decision to do so is the result of student input, either alone or as a result of the prelaboratory discussion.)

Cleaning Up

A very cursory introduction to solution concentration (molarity) is employed, since students will not have met the concept in lecture but must deal from the first day of laboratory with solutions whose labels give these mysterious *M* values.

Removal of the steel wool is best done with a boiling stick or bamboo barbecue skewer.

Answer to Prelaboratory Question

1.

Readings (mL)	Deviation
2.54	0.01
2.60	0.07
2.50	0.03
2.52	0.01
2.49	0.04

Average: 2.53
Average deviation: 0.05
Volume: 2.53 ± 0.05 mL

Data and Observations: Typical Results

1. Appearance of steel wool plus hydrochloric acid: Faint odor of HCl, bubbling, wool becomes discolored (red-brown).

   Appearance of steel wool plus acetic acid: Odor of vinegar; no visible changes.

2. Initial volume of water in reaction tube: 4.9 mL.
   Final volume of water in reaction tube: 4.1 mL.
   Total volume of reaction tube: 5.0 mL.
   Percentage decrease in volume in reaction in reaction tube: 16%.

   The assumption involved when unmarked tubes are used will result in a percentage decrease that is lower than the actual because of the rounded bottom of the tubes. Even so, the value should be close enough to give representative results, since the region where the water level is changing is cylindrical.

3. Mass of beaker, Erlenmeyer flask, culture tubes, and contents, weighted to the nearest milligram: 34.567 g.

   Appearance of solutions before mixing: Clear and colorless; no noticeable odors.

   Mass of beaker, Erlenmeyer flask, tubes and contents, weighted to the nearest milligram, after mixing calcium chloride, $CaCl_2$, with sodium bicarbonate, Na_2CO_3: 34.566 g.

   Appearance of solutions after mixing: White solid produced; at first, very little change, then milky, and finally began to settle.

   Mass change, if any: –0.001 g.

   (*Note:* Minor increases and decreases, say, ±2 to 3 mg, are normal.)

   Mass of beaker, Erlenmeyer flask, culture tubes and contents, weighted to the nearest milligram, after mixing calcium chloride, $CaCl_2$, with sodium carbonate, Na_2CO_3, and then sulfuric acid, H_2SO_4: 34.320 g.

   Appearance of solutions: White solid dissolves with evolution of gas; some spraying may be observed.

Analysis

1. Typical results appear above; if possible, make a data table on the board and have students enter their data. Presuming that all students use the same type of glassware, results should be similar. In any event, all should observe the same sorts of changes and note similar changes in total mass.
   a. Results are generally pretty consistent: ±1% to 2%. Variations here come from student experience and laboratory skills: speed with which the apparatus is assembled, accurate reading of volume levels, etc.
   b. Yes; Na_2CO_3, $NaHCO_3$
   c. Most students will recognize the "CO_3" parallel but will not make the connection with CO_2. Postlaboratory discussions are needed here.

2. Since all use the same type of tube, it is reasonable to expect at least some consistency, but expect random fluctuations due to both technique and interpretation of meniscus levels. Average deviation = sum of individual deviations from average ÷ number of students reporting. If unmarked tubes are used, this question might be left for the postlaboratory discussion.

Conclusions

1. Answers will vary; insist on interpretation *and* explanation for steps 1, 4, 5, and 6, per the blanks on the Report Sheet.

2. Answers will vary; require that a "no" be justified with reason(s) why no further judgments can be made.

3. Commonly, students say, "Do it again" or "Be more careful." Often you will get, "If I had known what to expect, I would have done it better or learned more," etc. Remind the student that science is an exploration; if repeating will help, repeat it! This is also a reason behind the urging to consult others and the use of group data.

Answers to Postlaboratory Questions

1. Since no heat is being supplied and the system is known to contain primarily water, the bubbling must be evidence of gas formation.

2. Student results.

3. Yes, but while rusting does occur, it is not always evident until students do the removal of the steel wool ball in the Cleaning Up section.

4. At this early point, such a question calls for speculation that most students are reluctant to risk. If they recall the gas evolution from Step 1, they may predict that the water

level will fall. In this case, proposing an answer and defending it are far more important than being "right."

5. Rusting requires a certain amount of moisture; this fact may or may not be familiar.

6 & 7. The goal is to encourage communication. A random check of those nearby should show that all saw the same changes and all recorded a slight mass loss due to mixing. While this loss probably will not be very reproducible, the fact that all found it means there was, in fact, some matter lost from the system.

8. a. The total mass of the reactants in a chemical reaction equals the total mass of the products in the reaction.
   b. One would not expect the same mass because two of the products of combustion were not weighed: the carbon dioxide and water that were formed.

EXPERIMENT

2 Densities of Organic Liquids

Placement: Early in the course; can be the second experiment in the first laboratory session or the first of two in the second (but see the Discussion Section of this in the Instructor's Guide for Experiment 5.)

Approximate time for completion: 30 minutes

Approximate preparation time: 20 minutes

Materials: Unless noted otherwise, amounts are per student, working individually

Apparatus

Milligram balances
Graduated syringe, 1 mL
Short (2 to 3 cm) length gum rubber or clear plastic tubing
1.0 x 0.01 mL Mohr (graduated) pipette, TC or TD
Sample vials with caps (4 per student)
Reaction tubes or 10 x 75 or 10 x 100 mm culture tubes (4 per student)
Safety goggles
Laboratory apron or coat

Reagents

Ethyl acetate, 50 mL
Hexane, 50 mL
Acetone, 50 mL
Unknown liquid, 2.5 mL per student—See "Notes on the Procedure"

Discussion

The technique of pipette use is introduced, along with the concepts of precision and accuracy and relative error (percentage error), while reinforcing the calculation of average deviation introduced in the preceding experiment.

Organic solvents offer such a wide enough range of densities that it is not necessary to resort to aqueous solutions of various concentrations. Their use does, however, mean that pipetting will be more of a challenge.

Our choice for the unknown is dichloromethane (assuming, of course, that the laboratory has adequate ventilation). While it is a halogenated hydrocarbon and so requires special disposal, it offers the unique feature of being more dense than water (hence reaffirming the identity). While not on the list, *tert*-butyl chloride (density = 0.842 g/mL) is another possibility, since it can be rendered disposal safe by converting to the alcohol. To the combined waste portions add a few drops of bromothymol blue, and then add 1 *M* NaOH dropwise with swirling until the color changes from yellow to blue.

Notes on the Procedure

Although we provide an explanation of the difference between TC and TD pipettes, you will need to show your students the particular type they are using and demonstrate use of the pipette, especially the technique for filling. While you may replace the syringe-tubing filling apparatus with standard pipette bulbs, we have found that there is far less problem with students overfilling the pipette and thus flooding the syringe than there was with bulbs. If you have disposable syringes fitted for Leur stopcock tips, these work fine; the 1-mL syringe from the Williamson Microscale kits sold by Kimble/Kontes will accept the tubing directly. Another factor favoring use of a 1-mL syringe is that it almost ensures that the volume used will be less than 1.00 mL, making it necessary for students to practice estimating volume readings and pay attention to significant figures.

The assumption is that there will be enough residual liquid in the pipette to carry out the miscibility test (which is also a test of relative density in some cases). Students will be tempted to draw a second sample of the liquid into the pipette for this; discourage them from doing so both in the interest of economy and for safety; less *is* better. The less they use, the less must be disposed of.

Answers to Prelaboratory Questions

1. 4°C: 6 significant figures; 25°C: 4 significant figures.

2. The gas (air) in the flask has mass. In the initial experiment, the air in the flask was "balanced by" the air surrounding the balance apparatus. When the entire system is placed in an evacuated chamber, this is no longer true. This will strike students as an "off the wall" sort of question, but it relates directly to the experiment, which is concerned with density, albeit of liquids, rather than of gases (air).

3. $0.750 - 0.735 = 0.015$; $0.015 \div 0.750 = 0.020 = 2.0\%$.

Answers to Postlaboratory Questions

1. a. Density 1: 0.894
Density 2: 0.908
Density 3: 0.910
Density 4: 0.890
Density 5: 0.892
Density 6: 0.902
Total: 5.396; average: 5.396 ÷ 6 = 0.899 g/mL
Deviations: 1: 0.005; 2: 0.009; 3: 0.011; 4: 0.009; 5: 0.007; 6: 0.003 g/mL

Total deviation: 0.044; average deviation: ±0.007 g/mL

b. Students answer; should be fairly close. In the preceding example, the average deviation was 0.007 against a mean value of 0.899; this corresponds to (0.007 ÷ 0.899 =) 0.8% error. Similarly, students should be within a percent or so of the accepted values. If they are not, technique may be faulty.

c. Average deviation gives information only about the precision of the experiment; it tells nothing about whether the value for the density is correct or not. On the other hand, relative percentage error *is* concerned with how closely the measured value matches the theoretical.

2. Since the volume is measured before evaporation, while mass is determined afterwards, evaporation would decrease the mass but not affect the volume. Density would be reduced.

3. First, the pipette is probably a more precise measuring instrument. From a more practical standpoint, however, cross-examination would be a problem if the same syringe were used (as the same pipette is used here) for all liquids. Further, the rubber seals of the syringe are likely to be attacked by some of the solvents used. Assuming that the pipette starts out well cleaned, each of the liquids used should be effectively gone before the next is added, so contamination should not be a problem if students follow the procedure given.

EXPERIMENT

3 Separation of a Mixture

Placement: Physical Properties; Separation of Mixtures

Approximate time for completion: 1.5 hours

Approximate preparation time: 20 minutes

Materials: 24 of each unless noted otherwise, assuming a class of 24 students working individually

Apparatus

Milligram balances (at least one for every 6 students; 1 for every 3 or 4 is better)
4 x 4 cm polyethylene bags (120; 5 per student) (Sandwich-size plastic bags may be substituted.)
Glazed or waxed paper squares
Magnets (8 to 12; can be shared)
Sublimation apparatus
- 25-mL filter flask
- Pipette bulb (as from a Pasteur pipette)
- Centrifuge tube with pluro adapter
- Hirsch funnel
- 12-mm (diameter) filter paper

Ring stand and clamp to support filter flask
Water aspirator or other source of vacuum (50-mL syringe will do)
Steam bath or hot plate with beaker of hot water or electrically heated sand bath
Watch glass (if option B is used in part 3 of the procedure)
Wooden applicator stick
100-mL beaker
10-mL Erlenmeyer flask
Filter paper (for collecting and drying polyethylene and said)

Reagents

Mixture of approximately the following composition:
- Sand, 25 g
- Iron filings (clean), 25 g
- Naphthalene, 5 g
- Polypropylene, isotactic (Aldrich 18,238-9) or finely shredded plastic milk container, 4 g. Note that the text states that polyethylene is used. This is not correct. Polyethylene has a lower melting point and often will melt during the sublimation process.
- Urea, 12.5 g

Ice, crushed or small pieces

Discussion

We have intentionally included sublimation in the procedure, believing it to be an important and underplayed concept and technique at the beginning level. For the same reason, we have placed emphasis on vaporization in the Prelaboratory Questions. Both solubility characteristics and relative densities are incorporated into the separation.

If you feel that you must omit sublimation (we hope not), you might replace the naphthalene with ammonium chloride. It will be lost during the evaporation of the aqueous layer due to decomposition: $NH_4Cl(aq) \rightarrow NH_3(g) + HCl\ (g)$. This presents students with an interesting problem, since the total mass recovered will be consistently less than 100%, and by a fairly reproducible amount (assuming that you encourage them to compare results). This suggests that something that was in the mixture was lost along the way.

Notes on the Procedure

Some parts of the experiment will be handled easily by the students, but your help both in the prelaboratory instruction and during the course of the experiment will be crucial for success in the sublimation and for the evaporation of water under reduced pressure. (The reduced-pressure option gives better results and does not include the risk of spillage.) Your help also will be needed for the evaporation procedure.

One critical point comes curing the sublimation. The mixture is best heated on a steam bath or boiling water bath; if the system is heated beyond about 100°C, the polypropylene softens a bit, forming a semisolid layer on the bottom of the filter flask. Over about 200°C, the entire contents remaining after sublimation fuses into a very difficult to handle mass. One of us (JGL) has had success having students use a light nonpolar solvent to partially dissolve and break up the mass. Some of the polypropylene is lost in this way, but the balance of the separation goes fine.

Answers to Prelaboratory Questions

1. Students should report densities, melting and boiling point data, as well as solubility information. Only urea is water soluble; this should lead to suggestions that urea could be isolated by adding water, filtering out the water-insoluble components, and then evaporating the water. This is exactly what they will do, although not as a starting point.

2. a. They seem to disappear.
   b. By their odor.
   c. The fact that they have an odor indicates they are at least partly in the vapor phase; therefore, they evaporate or, more correctly, sublime.

3. a. Yes; iron is an element.
   b. Sand is almost certainly not pure silicon dioxide, and the *poly* in polypropylene suggests an indefinite composition, hence a mixture.

Conclusions

1. To the first question, the answer is no. Magnetism is very selective, and sublimation works only for solids that have high vapor pressures (students won't use these words). If all components have similar solubility characteristics, filtration will not work too well. Students do not have the experience to know about fractional crystallization. Answers will vary to the second question. Many will mention the difficulty in getting solids out of containers and off the sublimation surface. Others will opt for the need to have everything dry for precise weighings.

Answers to Postlaboratory Questions

1. Adding water first (to dissolve urea) would impede the separation of iron by magnetism and would render the sublimation of naphthalene ineffective.

2. Sublime the naphthalene from the urea, or add water to selectively dissolve the urea.

3. On the basis of this, one would predict that both would float, since only halogenated hydrocarbons appear to be more dense than water. Interestingly, the *CRC Handbook* (73rd ed.) shows a value of 1.1536 g/mL for the density of naphthalene at 25°C but a much lower one (0.9625) at 100°C.

4. The separation of sand from gold and the separation of wheat from chaff are analogous to the separation of sand from polypropylene. They depend on the relative densities of the two substances being separated.

EXPERIMENT

4 Crystallization

Placement: Early, during discussion of pure substances and physical properties

Approximate time for completion: 30 minutes for part 1, 45 minutes for part 2

Approximate preparation time: 30 minutes

Materials: 24 of each unless noted otherwise, assuming a class of 24 students working individually

Apparatus

Electrically heated sand bath (at least 12)
10 x 100 mm reaction tubes, 48 pieces (2 per student)
Wood boiling stick
Hirsch funnel
12-mm (diameter) filter paper
25-mL filter flask
Pasteur pipette and bulb
Clamp and ring stand to support filter flask
Cotton or paper towel (for insulation in Step 3)
Filter paper for collecting recrystallized products, 48 pieces

Reagents

Impure phthalic acid, 25 g (mixed with enough methyl orange or similar dye to give a red-orange tint to the solution in water. This will be a *very* small quantity of the dye. The crystals will not appear to the student to be colored.
Aspirin tablets, 24 [1 tablet per student, 325-mg (5-grain) size]
tert-Butyl methyl ether, 50 mL
Hexane, 50 mL
Ice, for ice water bath

Discussion

The experiment involves two relatively straightforward purification steps, both of which can be completed in just over an hour. The phthalic acid is mixed with methyl orange to provide an obvious but easily removed contaminant.

The Prelaboratory and Postlaboratory Questions focus primarily on developing in students the ability to think critically about both the procedure and interpretation of the results and are intended to lay the groundwork for data analysis in later experiments.

Notes on the Procedure

1. No effort is made to determine the percentage recovery of the phthalic acid in part 1, but students are directed to attempt to determine the percentage of binder in their aspirin tablets.

2. It would be a good idea to demonstrate use of the Pasteur (or Beral-type) pipette for removal of the mother liquor from the tube containing the recrystallized phthalic acid. If plastic pipettes are to be used, the microtip style is better suited to this purpose than the ordinary thin-stem style. As the text suggests, gently expelling air while the pipette is being inserted into the tube will prevent solid from being withdrawn. Allow a few seconds for the crystals to resettle before drawing the liquid into the pipette. Here again, a slow, gentle release of pressure on the bulb ensures that only liquid, and not crystals, is pulled into the stem.

3. If available, hand lenses make for more detailed descriptions of the crystals produced in both parts of the experiment.

4. If *tert*-butyl methyl ether is not familiar to you, be aware that it does not present nearly the peroxide risk that is associated with other ether compounds, notably diethyl ether. Tertiary ethers do not peroxidize at all, and methyl ethers do so only under extreme circumstances. It may be of interest to note that its primary commercial use is as an additive in oxygenated automotive fuels.

5. While they are mentioned as an alternative heating source in Chapter 3, "Separation of a Mixture," this is the first experiment to call for use of the electrically heated sand baths, which are our heating device of choice in most of these experiments.

Answer to Prelaboratory Question

1. Sugar is the most likely. Others might include other drugs; any reasonable guess should be accepted.

Data and Observations

Recrystallization of Impure Phthalic Acid

Appearance of powder at the beginning of the experiment: Slightly pink or almost colorless.

Appearance of initial solution of powder: Pink or red-orange.

Effect of heating suspension of powder: It goes into solution.

Appearance of mixture after being cooled rapidly: Fine white crystals.

Appearance of mixture after being cooled slowly: Large clear crystals.

Which method probably gives purer crystals? Explain your choice: Slow cooling, smaller surface-to-volume ratio.

Color of crystallized phthalic acid: Colorless.

Color of filtrate: Pink.

Crystallization of Aspirin (Acetylsalicylic Acid)

Weight of aspirin tablet: Typically 325 mg.

Appearance of the initial solution of aspirin in the solvent: Cloudy (due to binder).

Appearance of filtrate: Should be clear.

Appearance of binder on the filter paper: A white powder.

Is supersaturation detected? Sometimes.

Appearance of crystals of aspirin: Clear glistening blades.

Weight of dry recovered aspirin: Student answers; probably about 250 to 300 mg; product may be wet, therefore heavy.

Percentage recovery of recrystallized aspirin: Student answers.

Answers to Postlaboratory Questions

1. A saturated solution is one in which no more of the solute will dissolve at the given temperature. Depending on the substance, the solution may be concentrated, which means that many grams of solute per 100 mL of solvent are dissolved, or it may be dilute, which means that few grams of solute per 100 mL of solvent are dissolved.

2. It is assumed that the impurities are soluble. If they were not, they would simply be removed by filtration.

3. Commercially, any substance that is sold as crystals probably has been purified by crystallization. These include the flavor enhancer MSG (monosodium glutamate), sugar, salt, a variety of fertilizers (sodium nitrate, potassium nitrate, and ammonium nitrate), and calcium chloride (used in deicing roads).

EXPERIMENT

5 Sublimation

Placement: Early in course, with discussions of physical separation of mixtures. Can also be inserted *before* Experiment 2, "Separation of Mixtures."

Approximate time for completion: 30 minutes

Approximate preparation time: 35 minutes (including preparation of mixture)

Materials: 24 of each, except as noted, assuming a class of 24 students working individually

Apparatus

Milligram balances
Wax paper squares for weighing impure and purified samples (48)
25-mL filter flask
Pluro stopper for tube
15-mL glass centrifuge tube
Rubber pipette bulb
Electrically heated sand bath
Hair dryer (optional, but helpful) (pistol type) or heat gun
Spatula
Safety goggles

Reagents

Unknowns: About 500 mg of each. Each unknown should be mixed with a small quantity of decolorizing charcoal (activated carbon), about 1 to 2 mg of charcoal per 500 mg of unknown. This should give a dark black color. If necessary add more charcoal.
    Naphthalene
    p-Dichlorobenzene
    Benzoic acid
Ice

Discussion

This experiment introduces a topic that always appears in texts but is rarely treated in the laboratory portion of general chemistry courses. In fact, students use the technique in Chapter 2, "Separation of Mixtures." You may choose, therefore, to insert between Chapters 1 and 2 for this reason.

Efforts have been made, both through the introduction and with the Prelaboratory and Postlaboratory Questions, to emphasize that sublimation is a familiar process to virtually everyone, even though it may not seem so.

The procedure is only semiquantitative in design; the use of approximate melting ranges is designed to show students that it is quite possible to make some definitive judgments without having precise measurements from which to work. When you do not have all the data you would like, the next best thing is to make full use of what you do have.

The specific choices of unknowns were based on a desire to have familiar substances that we assumed would be readily available at most schools and ones that sublime at normal atmospheric pressures.

Notes on the Procedure

1. Demonstrate assembly of the sublimation apparatus. It is best if the pluro stoppers are put on the centrifuge tubes by the instructor or laboratory assistants rather than being left for the individual students. Once mounted on the centrifuge tubes, they can be left there permanently.

2. Spend some time during the prelaboratory discussion with the steps involved in removing the sublimed product, especially the need to bring the tube to room temperature before removal from the flask, since atmospheric moisture will condense on a cold tube, wetting the product if the cold water is not replaced.

Answers to Prelaboratory Questions

1. This calls on students to iterate the procedure.

2. Solid room air fresheners and deodorant "cakes" used in public restrooms—which are made from one of the unknowns—are good examples.

3. As the name implies, the food is frozen, and then water in the food is sublimed under high vacuum. The temperature of the food never rises above 0°C.

Answers to Postlaboratory Questions

1. Yes. the unknown was dark black to begin with. Sublimation gave pure white crystals with a sharp melting point.

2. Some of the product will sublime onto the walls of the flask. Unless the flask surface is kept warm, this is almost unavoidable.

EXPERIMENT

6 Solubility and Solutions

Placement: Either with early chapter on mixtures or in connection with the more extended discussion of solutions in a later chapter

Approximate time for completion: About 1 hour

Approximate preparation time: 1 hour

Materials: 24 of each, except as noted, assuming a class of 24 students working individually

Apparatus

24-well test plates
Pasteur or plastic Beral pipettes (48 plus the ones containing reagents)
Conductivity tester
Stirrers, glass or plastic (melting point capillaries or toothpicks)
Micro vials, polypropylene or glass (18 or 24, if possible, but may be cleaned periodically if fewer are used)

Reagents

Amounts are for a class of 24 students working alone. Directions for preparation may be for more than the amount needed for one class; convenience has been taken into account.

Solids: About 5 g each

Part 1

Sodium bicarbonate
Copper(II) nitrate trihydrate
Potassium carbonate
Calcium carbonate

Nickel(II) nitrate hexahydrate
Sulfur
Sodium chloride
Calcium chloride
Ammonium chloride
Silver chloride
Zinc oxide
Calcium oxide
Silicon dioxide
Sodium carbonate
Calcium sulfate
Copper(II) sulfate pentahydrate
Potassium permanganate
Aluminum powder
Zinc powder
Tin powder
Carbon (charcoal, powder)
Magnesium, turnings or powder
Iron (filings, powder)
Sucrose

Part 2

Cholesterol
Naphthalene
p-Dichlorobenzene
Polystyrene*
Benzoic acid
Caffeine
Citric acid
Iodine
Vanillin

Liquids and solutions: About 10 mL of each will be needed; place pipettes or dropper bottles in convenient locations.
Distilled water, about 50 mL
6 *M* HCl(*aq*) (*Slowly and with constant stirring,* add 50 mL of concentrated acid to 50 mL of distilled water. Allow to cool before use.)
6 *M* NaOH(*aq*) (*Slowly and with constant stirring,* add 24 g of NaOH pellets to 50 mL of water. Much heat will be generated. Dilute the solution to 100 mL. Allow it to cool before use.)
Methanol
2-Propanol
Ethylene glycol
Acetic acid, glacial
Cyclohexane
tert-Butyl methyl ether
Ethyl acetate
Acetone

Discussion

This is a highly materials-intensive exercise, and students are left with a sizable amount of data to sift through in the search for patterns. We see this as both instructive and worthwhile, but you can certainly pare down the list of reagents. As noted under "Materials," only small amounts of the various liquid reagents are needed, but it is most convenient, and in the long-run frugal, to put out snap-cap dropper bottles, placed at accessible locations about the laboratory. This means that you will put out 10 or 20 times the needed volumes, but there should be less waste and risk of contamination this way than if you were to have students withdraw their own small samples from a single stock container.

* Polystyrene powder may be purchased, but a tiny speck of a white disposable expanded polystyrene cup will do fine.

Similarly, the solids are most easily dispensed from small glass or plastic containers, such as the 1- or 2-oz "soufflé cups" available from grocery or restaurant supply stores. These cups have the added advantage of available snap-on lids, which are especially useful for naphthalene and *p*-dichlorobenzene.

Notes on the Procedure

There is really nothing potentially confusing here for students. Once you go through a brief prelaboratory discussion, mainly detailing the location of reagents, you need to make sure students understand that the focus is on establishing patterns of behavior that may be connected with other properties, appearance, or names of the various solutes. From there on, the experiment should proceed smoothly.

Answers to Prelaboratory Questions

1. a. $CaCO_3$
   b. CaO
   c. $CaSO_4$
   d. $NaCl$
   e. $C_{12}H_{22}O_{11}$
   f. $NaHCO_3$
   g. Many students will return to baking soda, $NaHCO_3$. Other possibilities are washing soda, sal soda, and soda ash, all forms of sodium carbonate, Na_2CO_3.
   h. SiO_2

2. b. Copper(II) nitrate trihydrate: $Cu(NO_3)_2{\cdot}3H_2O$
   c. Nickel(II) nitrate hexahydrate: $Ni(NO_3)_2{\cdot}6H_2O$
   f. Calcium chloride: $CaCl_2$
   g. Ammonium chloride: NH_4Cl
   h. Silver chloride: $AgCl$
   i. Zinc oxide: ZnO
   l. Sodium carbonate: Na_2CO_3
   m. Potassium carbonate: K_2CO_3
   p. Copper(II) sulfate pentahydrate: $CuSO_4{\cdot}5H_2O$
   q. Potassium permanganate: $KMnO_4$

3. a. Citric acid
   b. Caffeine
   c. Vanillin
   d. Ethylene glycol
   e. Methanol
   f. Iodine
   g. Acetone
   h. *tert*-Butyl methyl ether
   i. Naphthalene (or *p*-dichlorobenzene)
   j. Sucrose

k. 2-Propanol
l. Acetic acid
m. Cholesterol
n. Benzoic acid

Answers to Postlaboratory Questions

1. Only a few of these would normally be assigned.

2. Ionic substances conduct an electric current and also, in general, dissolve in water. Covalent substances do not conduct an electric current and dissolve in nonpolar solvents. For organic compounds, water solubility is observed most often in cases where the structure includes -OH or -COOH groups. (Acetone, while very soluble, is an exception.)

EXPERIMENT

7 Isolation of the Silver in a Dime

Placement: With the chapter on percentage composition

Approximate time for completion: 2.5 hours

Approximate preparation time: 40 minutes

Materials: 24 of each unless noted otherwise, assuming a class of 24 students working individually

Apparatus

Milligram balance (at least one for every 12 students)
50-mL beaker
Watch glass, 2 in. (5 cm) diameter
Graduated cylinder, 25 or 50 mL (one for each group of five students)
10-mL Erlenmeyer flask
Stirring rod
Steam bath or hot plate adjusted to 95°C

Reagents

Amounts are for a class of 24 students working alone. Directions for preparation may be for more than the amount needed for one class; convenience has been taken into account.

Pre-1965 dime (one per five students), purchased from coin store for approximately $0.40.
Nitric acid, HNO_3, 75 mL concentrated (15.8 *M*)
Potassium chloride solution, KCl, 100 mL, 2*M* (Dissolve 15 g of solid KCl in 100 mL of distilled water; this concentration is not critical.
Nitric acid, HNO_3, 500 mL, o.3 *M* (*Slowly and with constant stirring,* add 10 mL of concentrated acid to 490 mL of distilled water)

Ammonia, $NH_3(aq)$, 15 mL concentrated aqueous. *Caution: Strong fumes; use only in hood.*
Acetone, 500 mL (wash bottles or similar)
Sodium hydroxide solution, NaOH, 75 mL 6 *M* (*Slowly and with constant stirring*, dissolve 24 g of solid NaOH pellets in 100 mL of distilled water. *Caution: The mixture will become quite hot;* allow to cool before using.)
Sucrose (ordinary table sugar), 24 g
Baking soda (for neutralization during cleanup)
Sodium carbonate, $Na_2CO_2(s)$ (flux in optional experiment)

Discussion

This experiment has been adapted from an experiment performed at Harvard University for at least the last 40 years. Except for the necessity of an occasional visit to a coin shop, the use of silver dimes is not a problem. The coin dealer will have a number of dimes that, because of wear and the large number minted, have no particular value beyond the silver they contain. Recently, dimes were selling at $0.38 each, which was almost exactly the value of the silver in them. Each student uses one-fifth of the dissolved dime, so the experiment is not particularly expensive.

This is a highly practical example of percentage composition (although it involves a mixture rather than a compound) and incorporates a number of techniques that students will need in later experiments.

The chemistry involved is basically the same as that used in the popular silvered flask demonstrations using Tollen's reagent; the sodium hydroxide hydrolyzes the sucrose to simple reducing sugars.

If you are equipped for it, the optional formation of the silver bead is well worthwhile. If nothing else, consider it as a demonstration using the combined products of two or three students.

Notes on the Procedure

Procedure Overview

Warming of the beaker with the dime, water, and nitric acid will almost certainly be necessary. Once the reaction is going vigorously, heating may be discontinued, since the reaction is quite exothermic. The dime will dissolve in about 4 to 5 minutes with copious evolution of nitrogen dioxide. This part of the experiment must be carried out in a good hood. On cooling to room temperature, the blue solution deposits crystals of silver nitrate, so an additional 4 mL of water must be added to the solution.

As a group, the students carry out this part of the experiment and calculate the mass of one-fifth of the solution and weigh this amount into a tared Erlenmeyer flask. It requires about 4 mL of 2 *M* potassium chloride to precipitate all the silver as the chloride.

The flask containing the curdy white precipitate is heated with stirring on a steam bath or a hot plate set to just under the boiling point of water. Boiling the solution on a hot plate often causes the solution to bump with loss of product. After about 10 minutes of heating and stirring on the steam bath, the supernatent is clear. The speed with which the solution clarifies is proportional to the amount of stirring.

It is remarkably easy to pour off the aqueous layer from the heavy precipitate without losing any particles of silver chloride. The precipitate is washed four times with 5-mL portions of dilute nitric acid and then three times with 3-mL portions of acetone. Cautious warming on the steam bath will drive off most of the acetone without the mixture bumping. The flask is then placed in a labeled beaker and heated in an oven for 30 minutes at 110°C. The principal source of error in this experiment seems to be inadequate drying of the precipitate. If the calculated percentage of silver is too high, the students should dry the silver chloride again before proceeding with the rest of the experiment.

The silver chloride is reduced to elemental silver using sucrose. The reaction mixture is heated on the steam bath until the supernatant is clear and brown, not cloudy brown. Most of the silver forms in fairly large particles, so decanting the supernatant is effective, but not as effective as decanting from silver chloride, because some of the silver is in the form of a very fine powder that is lost in decantation. Because of this loss, the calculated silver content of a dime was found for five trials to be 87.64 ± 0.49%. The calculated amount of silver, based on the silver chloride present, was 89.98 ± 0.15% silver (measurements were made to four significant figures).

The coin dealer from whom our dimes were purchased sold all dimes, regardless of weight, for $0.38 each. We chose a relatively unworn dime that weighed more than the average, so the silver in the dime was worth exactly what we paid for it. Older, worn dimes would, of course, contain less silver, and the coin dealer would make a profit on those. Coin silver is 90.00% silver and 10.00% copper.

Additional Notes

1. Stress the safety information, in particular the hazards associated with NO and NO_2 generated during dissolving of the dimes. Students should not remove the beaker from the hood nor carry out the division of the mixture until the reaction has ceased and the mixture has cooled to room temperature (5 to 10 minutes).

2. Most of the reagents that are to be used by individuals are best dispensed from small snap-cap plastic bottles (2 *M* KCl, 0.3 *M* HNO_3, acetone, 6 *M* NaOH). Because of its odor, the concentrated ammonia should be placed in the hood.

3. The very dilute (0.3 *M*) nitric acid used to rinse the precipitate of AgCl serves to inhibit photo-oxidation of the solid.

4. In the prelaboratory, stress the fact that during the rinsing/decanting parts of the procedure, it is better to leave some liquid behind than to lose any of the solid. In cases where the solid does not settle well, students may want to use a microtip plastic pipet to draw off the liquid. This might be especially helpful for removal of the last acetone rinse of the silver chloride prior to warming and evaporation to constant mass.

5. The "warm" water used to rinse the reduced silver should be over 40°C but need not be much hotter than that. If students must heat their own water, no thermometers are needed; if the water is hot enough to steam the sides of the beaker, it is hot enough to use here.

6. If you have students make the silver button (optional experiment), be sure they know how to handle the gas-oxygen torch. It is best if you or a student who is experienced with the apparatus either does all the torch work or closely oversees the process.

Answers to Prelaboratory Questions

1. Light causes it to air oxidize. This is the basis for the photographic process.

2. 3.696 g AgBr x (107.9 g Ag/187.8 g AgBr) = 2.123$_5$ g theoretical yield of Ag (use of the subscript indicates that the 5 is not a significant digit). (2.000 g Ag ÷ 2.123$_5$ g Ag) x 100% = 94.18% yield.

Data and Calculations

These data are from an experiment carried out using an analytical balance.
Mass of beaker: 27.6852 g
Mass of dime: 2.4976 g
Mass of beaker, dissolved dime, and aqueous solution: 61.9442 g
Mass of solution: 34.2871 g
Mass of Erlenmeyer flask plus about one-fifth of dime solution: 22.2804 g
Mass of Erlenmeyer flask: 15.4286 g
Mass of solution: 6.8518 g
Fraction of entire solution in flask: 6.8518 g/34.2871 g
Mass of flask plus silver chloride: 16.0207 g
Mass of flask: 15.4286 g
Mass of silver chloride: 0.5984 g
Atomic weight of silver: 107.87
Molecular weight of silver chloride: 143.32
Fraction of silver in silver chloride: 0.75265
Calculated mass of silver in your silver chloride: 0.45039 g
Percent of silver in dime from AgCl mass: 90.23%
Mass of silver obtained by reduction of AgCl: 0.4383 g
Calculated mass of silver in dime from mass of silver in aliquot: 2.1933 g
Percent of silver in dime from silver mass: 87.82%
Cost of dime from coin dealer: $0.38
Current price of silver per gram (The price is usually given in Troy ounces; one Troy oz = 31.04 g.): $0.1707
Value of silver in your dime: $0.3817

Answers to Postlaboratory Questions

1. Some material may be lost during division of the original dime solution. Insufficient coagulation of the precipitate will mean loss of AgCl. Exposure to light may convert some of the silver chloride to silver oxide, which will not react as easily with the reducing sugar.

2. In both reactions the silver is the reducing agent and the nitric acid is the oxidizing agent. The silver is oxidized to silver ion, and the nitric acid is reduced to NO and NO_2.

EXPERIMENT

8 Synthesis of Manganese(II) Chloride

Placement: Percentage composition of a compound, empirical formula

Approximate time for completion: 1.5 hours

Approximate preparation time: 30 minutes

Materials: 24 of each unless noted otherwise, assuming a class of 24 students working individually

Apparatus

Milligram balances (one for every eight students)
Electrically heated sand bath
10-mL Erlenmeyer flask or 10 x 75 mm culture tube
Desiccator
Forceps, for handling hot flask

Reagents

Amounts are for a class of 24 students working alone. Directions for preparation may be for more than the amount needed for one class; convenience has been taken into account.

Manganese metal, 2.5 g, crushed into small (50- to 100-mg) pieces
Hydrochloric acid, 10 *M*, 25 mL (Add 42 mL of concentrated acid to 8 mL of distilled water. *Caution: HCl vapors are highly corrosive;* work in the fume hood. Wear safety goggles and apron.)

Discussion

This experiment goes quickly and smoothly. We have written it as a percentage yield experiment, but it also can be done as an empirical formula determination. In the latter case, students first determine the mass of chlorine in the product, then convert the masses of manganese and chlorine to moles, and finally determine the mole ratio and the formula. Typically, the Cl/Mn ratio is found to be about 2.1:1 or 2.2:1.

Percentage yields generally run on the high side, mostly due to the rapid rehydration of the product.

Notes on the Procedure

1. Stress the safety information, in particular the hazards associated with HCl fumes (this is mentioned in the caution statements in the student version but bears repeating).

2. Occasionally, manganese chips will have a coating of manganese sulfide, which results in generation of highly toxic hydrogen sulfide gas. If the odor of H_2S is detected, students should leave the 10-mL Erlenmeyer flask (labeled for identification) in the hood until the reaction is complete. If possible, the heating should at least be started in the hood, too, since H_2S has a significant solubility in water.

3. *Equipment substitutions:*
   a. In the absence of electrically heated sand baths, food tins (especially the large-size tuna fish cans) will do. The can should not be filled to greater than half its total depth with sand.
   b. In addition, 100-mm test tubes may be used in place of the Erlenmeyer flasks, but spattering is likely to be a problem, especially at the beginning. This can be controlled somewhat by use of a small carborundum boiling chip, but some loss due to spattering is unavoidable, as is significant refluxing.

4. Reasonable results are obtained even without the repeated heating and desiccation. If the flask and contents are heated according to the procedure and the product is almost white when flask and contents are desiccated, the percentage yield will be within acceptable limits (95% to 110%).

5. Assuming the procedure is followed, the single most common source of experimental error is rehydration of the product.

Answer to Prelaboratory Question

1. The mass of chlorine in the product is the difference between the mass of product and the initial mass of manganese. Dividing the mass of each element by its atomic mass gives the number of moles of each. The ratio of moles of chlorine to moles of manganese must equal the atom ratio for the compound—presumably 2:1. When the experiment is done for this purpose, most students come out with a ratio of 2.1:1 or 2.2:1.

Data and Calculations

Mass of flask and boiling stone: 15.740 g
Mass of flask, stone, and manganese: 15.839 g
Mass of manganese used: 0.099 g
Moles of manganese used: 0.0018 mol
Mass of flask and boiling stone: 15.740 g
Mass of flask, stone, and manganese chloride: 15.973 g
Mass of manganese chloride prepared: 0.233 g
Moles of manganese chloride prepared: 0.0019 mol
Theoretical yield of $MnCl_2$ in grams: 0.23 g (2 significant figures)
Your percent yield of $MnCl_2$: 100%

Conclusions

1. a. The mass of product will include the acid solution and so will be too high, as will the percentage yield.
   b. The yield will be too low. This is extremely unlikely; $MnCl_2$ melts at 650°C and boils at 1190°C.
   c. Moisture will be reabsorbed, changing $MnCl_2$ to the dihydrate; the mass and percentage yield will increase. This is the most likely error in the experiment.

2. Greater; manganese(II) chloride is stable to several hundred degrees, and incomplete digestion should not be a problem for the student who is conscientious. All other common errors, most notably incomplete evaporation and rehydration, result in high percentages.

3. On skin or clothes: Remove affected clothing; flush with copious amounts of water. (This is no time for modesty.) On desk top: Neutralize with $NaHCO_3$ (baking soda) and then swab up with wet towels. In both cases, the instructor must be notified immediately and his or her instructions followed explicitly.

4. a. $H_2SO_4 + Mn \rightarrow MnSo_4 + H_2$
   b. Account would need to be taken of the change in mass and thus of the numbers of moles of product formed. Its formula weight has increased. The number of moles of hydrogen released is the same as if HCl had been used.

EXPERIMENT

9 Analysis of Copper Oxide

Placement: Percentage composition of a compound, empirical formula

Approximate time for completion: 1 hour

Approximate preparation time: 20 minutes

Materials: 24 of each unless noted otherwise, assuming a class of 24 students working individually

Apparatus

Milligram balances (one for every eight students)
20- or 30-mL beaker
Pasteur pipette or thin-stem Beral-type pipette
Steam bath; students can share (see "Notes on the Procedure" below)
Drying oven
Desiccator

Reagents

Amounts are for a class of 24 students working alone. Directions for preparation may be for more than the amount needed for one class; convenience has been taken into account.

Copper(I) or copper(II) oxide, 6 to 8 g (see "Notes on the Procedure" below)
Hydrochloric acid, 6 *M*, 60 mL (Add 25 mL of concentrated acid to 50 mL of distilled water. *Caution: HCl vapors are highly corrosive;* work in the fume hood. Wear safety goggles and apron.)
Distilled water (in wash bottles)
Aluminum metal, strip or wire, about 1 g

Discussion

This experiment introduces some of the terminology of redox reactions but does not dwell on the concepts, since it is intended for placement somewhat earlier in the course. When carried out with copper(II) oxide, the reaction goes quite smoothly, and even students with somewhat questionable technique are able to get good results. Two important types of chemical reactions are involved; (1) the reaction between metal oxide and an acid to produce water and the metal salt of the acid, and (2) the single replacement reaction between a more active metal and a compound of a less active one.

Notes on the Procedure

1. While the student is told that the original oxide may be either copper(I) or copper (II), the very low solubility of copper(I) chloride, along with its ease of oxidation to copper(II), makes copper(I) oxide a less attractive choice.

2. A steam bath may be made by clamping the 10-mL Erlenmeyer flask at a height 2 to 3 cm above the mouth of a 250-mL Erlenmeyer flask in which water is being boiled. As noted earlier, two students can conveniently share the same steam bath.

3. Small containers (such as plastic soufflé cups) of the unknown oxide may be placed near the balances. This minimizes transporting the fine powder. The acid is conveniently dispensed from snap-cap polyethylene dropper bottles.

4. The aluminum pieces should be big enough that easily recognized bits of unreacted metal can be removed efficiently when replacement of copper by aluminum is complete. Aluminum powder or granules are not as satisfactory, unless an excess of acid is maintained to dissolve any unreacted metal.

5. *Substitutions:* Aluminum forms an extremely hard oxide coating, so it should be cleaned before use. Zinc strips have been tried, but with poor success, perhaps due to an oil covering that was not removed properly before use. In any event, the use of aluminum requires students to deal with a mol ratio other than 1:1.

Answer to Prelaboratory Questions

1. $CuO + 2HCl \rightarrow CuCl_2 + H_2O$
   $2Al + 3CuCl_2 \rightarrow 2AlCl_3 + 3Cu$
   CuO: $(63.54 \div 79.54) \times 100\% = 79.99\%$ Cu
   Cu_2O: $(127.08 \div 143.08) \times 100\% = 88.82\%$ Cu

Data and Calculations

Student data. The results are ±3% to 4% Cu for copper (II) oxide, and ±5% to 8% for the less soluble copper(I) oxide.

Conclusions

1. As noted earlier, CuO gives very good results, Cu_2O not as good. Even so, there should be no trouble with this, since the percentages of copper are distinctly different.

2. Copper(II) oxide (or cupric oxide); copper(II) chloride (cupric chloride); $CuCl_2$.
Copper(I) oxide (or cuprous oxide); copper(I) chloride (cuprous chloride); CuCl

Answers to Postlaboratory Questions

1. (0.30 g Cu_2O ÷ 143.08 g/mol) x (2HCl/1 Cu_2O) = 0.0042 mol HCl; 0.0042 mol HCl x 1000 mL/6 mol HCl = 0.70 mL

2. When copper is removed from the metal strip, no new coating forms if the copper has all been replaced, and the blue color of the solution (due to copper(II) ion) is replaced by a grayish tinge.

3. a. 2.5 mL x 6 *M* mmol = 0.015 mol
   b. 0.015 mole of acid; 0.3 g/148.08 g/mol of Cu_2O = 0.002 mole; ratio 7.5:1. 0.015 mole of acid, 0.3 g/79.54 g/mol of CuO = 0.0038 mole; ratio 4:1.
   c. The oxide is the limiting reagent.

EXPERIMENT

10 Gravimetric Determination of Chloride Ion Concentration

Placement: Stoichiometry

Approximate time for completion: 1.5 hours

Approximate preparation time: 30 minutes

Materials: 24 of each unless noted otherwise, assuming a class of 24 students working individually

Apparatus

Milligram or analytical balance (at least one for every 6 students)
Pasteur pipette
Cotton for filters; 2 to 3 ordinary cotton balls are enough for one class
Wood applicator sticks (Fisher) (or small wood cooking skewers)
10 x 75 mm culture tube or test tube
Centrifuges
Drying oven (105 to 110°C)
Desiccator(s)
1-mL volumetric pipette, calibrated pipette, or syringe with Eppendorf tip (for solution unknowns only)

Reagents

Amounts are for a class of 24 students working alone. Directions for preparation may be for more than the amount needed for one class; convenience has been taken into account.

Nitric acid, 10 mL, 8 *M* (Add 25 mL of concentrated acid to 25 mL of distilled water. *Caution: HNO_3 vapors are highly corrosive;* work in the fume hood. Wear safety goggles and apron. The solution will get slightly warm; allow to cool before using.)

Distilled water (in wash bottles)
Silver nitrate solution, 25 mL, 1.00 *M* $AgNO_3(aq)$ (Dissolve 8.5 g of solid reagent in 50 mL of distilled water. Store in amber glass bottle or cover bottle with foil.)
Acetone, 15 mL
Unknown mixture(s), about 10 to 15 g, depending on balances used

Discussion

This is a familiar experiment, although not on this small scale. In writing the procedure, and especially the questions that accompany it, we have tried to lead the students in such a way that their understanding of the connection between moles and mass is heightened.

Notes on the Procedure

1. The choice of tube size is important. The calibrated 10 x 100 mm reaction tubes that we prefer for many experiments will not work here due to the narrow diameter. Some 10-mm culture tubes are sufficiently wide to accept the Pasteur pipette "vacuum cleaner" device, but we strongly recommend a 10 x 75 mm tube, which is easier for most students to manipulate.

2. The silver chloride precipitate will turn gray while students work. This is unavoidable but should make little difference in the ultimate result, assuming students maintain a good working pace.

3. The choice of solution or solid unknown is up to you. Solutions are easier for students to deal with, assuming that they are comfortable with volumetric pipettes, and you are more assured of uniform mixing with solutions. Solid unknowns provide students with practice using the balance, and many students have an easier time with them, although results may be less uniform.

4. Students tend to get frustrated with pulling the precipitate up into the filter pipette. It will help if you are available to provide an extra pair of hands at least for the first ones; they can help the others while they wait for their pipettes to cool from the oven.

5. *Substitutions:* We have tried 10-mL flasks and 20-mL beakers in place of the 10 x 75 mm tubes. The only drawback was that students wanted to set the beaker or flask on the desktop while vacuuming out the AgCl. This led to spills and loss of product.

Answers to Prelaboratory Questions

1. $Ag^+(aq) + Cl^-(aq) \rightarrow AgCl(s)$

2. 0.100 g x (1 mol $BaCl_2 \cdot 2H_2O$/244.2 g $BaCl_2 \cdot 2H_2O$) x (2 mol AgCl/1 mol $BaCl_2 \cdot 2H_2O$ x (143.4 g AgCl/mol AgCl) = 0.117 g AgCl

3. 0.100 g x (1 mol $BaCl_2 \cdot 2H_2O$/244.2 g $BaCl_2 \cdot 2H_2O$) x (2 mol $AgNO_3$/1 mol $BaCl_2 \cdot 2H_2O$) x 1.00 L/1.00 mol $AgNO_3$ = 8.19×10^{-4} L = 0.819 mL

4. Use of a large excess of silver nitrate solution ensures complete precipitation of the chloride. AgCl is sufficiently low in solubility that there is little concern that excess solution will redissolve some of the precipitate.

   This experiment could be used to determine the number of chloride ions bound to manganese. From the mass of Mn, the moles of Mn can be calculated. From the mass of AgCl, the number of moles of AgCl and hence of Cl can be calculated. The ratio of the number of moles of Cl to Mn should be a small whole number (2).

Data and Calculations

Student data.

Conclusions

1. The formula for silver chloride is AgCl: one chloride per formula unit. Assuming that all the chloride from the unknown becomes silver chloride, the moles of chloride ions in the unknown must equal the moles of AgCl precipitated.

2. Like silver chloride, KCl is a 1:1 electrolyte. Barium chloride ($BaCl_2$), however, has two chloride ions per formula unit, so mol Cl^- = 2 x mol $BaCl_2$

3. Student data, using the equations supplied above.

Answers to Postlaboratory Questions

1. a. Not all the chloride is precipitated; total moles of chloride is too low.
   b. The apparent mass of silver chloride will be greater due to presence of water. This will make the total mass m and the total moles of chloride t both larger than they should be. Since mol Cl^- from $BaCl_2$ = 2 mol $BaCl_2$, the percentage of KCl will come out larger than it should. (Students are unlikely to recognize the last part, but they should see that t and m will both be too large.)
   c. Warm objects appear lighter than they really are due to convection currents. This is the reverse of (b): t and m will both be too small, and the percentage of barium chloride in the mixture is raised.

2. Student responses; remind students that it is their own experiment, not just general difficulties, that is to be addressed.

3. Since one of the goals is to determine the percentage of chloride ion in a mixture, addition of hydrochloric acid, itself a source of chloride, would defeat the purpose.

4. 0.0716 g AgCl/143.4 g AgCl/mol AgCl = 0.0005 mol AgCl and thus the same number of moles of Cl in the unknown. 0.0212 g unknown/0.0005 mol Cl = 42.4 g/mol formula weight of unknown if it contains one Cl, 84.8 if it contains two, 127.2 if it contains three.

EXPERIMENT

11 Gravimetric Determination of Sulfate Ion Content

Placement: Stoichiometry

Approximate time for completion: 1.5 hours

Approximate preparation time: 30 minutes

Materials: 24 of each unless noted otherwise, assuming a class of 24 students working individually

Apparatus

Milligram or analytical balance (at least one for every 6 students)
100-mm test tube or reaction tube
Electrically heated sand bath or boiling water bath (1 per 2 students is sufficient)
Desiccator
Centrifuge

Reagents

Amounts are for a class of 24 students working alone. Directions for preparation may be for more than the amount needed for one class; convenience has been taken into account.

Solid unknown(s), 5 g (See "Notes on the Procedure" below)
Hydrochloric acid, 1 *M*, 10 mL (Add 8 mL of concentrated acid to 90 mL of distilled water. The exact concentration is not critical. *Caution: HCl vapors are highly corrosive; work in the fume hood.* Wear safety goggles and apron. The solution will get slightly warm; allow to cool before using.)
Barium chloride, 25 mL, 1.0 *M* (Dissolve 24.4 g of the solid dihydrate, $BaCl_2 \cdot 2H_2O$, or 20.8 g of the anhydrous solid in 100 mL of distilled water. *Caution:* $BaCl_2$ *is toxic; wash your hands thoroughly after handling.*)

Acetone (for rinsing), 25 mL (Wash bottles are the most convenient dispensers; while you will put more acetone out, they will not be wasteful.)
Distilled water (in wash bottles)

Discussion

This introductory quantitative analysis experiment can be completed easily by students. The fact that it involves one of the relatively few low-solubility sulfates helps reinforce the exceptions in solubility rules for students.

Notes on the Procedure

1. The logical choice for unknowns are alkali metal sulfates (Na_2SO_4, K_2SO_4, etc.), although others can be used. An option is to use various mixtures of a soluble sulfate and an "inert" contaminant (e.g., Na_2SO_4 and NaCl). Alternatively, you can draw on the implied suggestion of Prelaboratory Question 4 and give students hydrated sulfates. If the identity of the cation is supplied, students should be able to determine the number of waters of hydration.

2. The choice of tube size is of no particular importance. The calibrated 10 x 100 mm reaction tubes that we prefer for many experiments will serve, as will almost any tube that will fit in your centrifuge. Centrifuge tubes, if available, are an excellent choice, but the experiment works fine without. The very thin-walled disposable tubes sometimes break if the centrifuge is too powerful. Check a few before deciding what sort of tube to use.

3. Emphasize the need to use a minimum volume of distilled water in Part 1 of the procedure.

4. To reduce the risk of dissolving the product during rinsing (Part 3), have students chill their distilled water wash bottles before and during that part of the experiment.

5. Make sure that students understand that they are trying for a constant mass of precipitate so that the ±0.5% refers to the mass of just the solid and does not include the tube.

6. Check local regulations regarding disposal of the barium sulfate. If any is wet (it should not be), allow it to dry thoroughly before marking for removal.

Answers to Prelaboratory Questions

1. $Ba^{2+}(aq) + SO_4^{2-}(aq) \rightarrow BaSO_4(s)$

2. To ensure complete precipitation of the sulfate ion.

3. 0.0676 g $BaSO_4$ x (1 mol $BaSO_4$/233.4 g $BaSO_4$) x (1 mol SO_4^{2-}/1 mol $BaSO_4$) x (96.1 g SO_4^{2-}/mol) = 0.02783 g SO_4^{2-}. 0.02783 g/0.1000 g) x 100% = 27.83%; the fourth significant figure is not truly significant, since the mass of the precipitate is given only to three significant figures.

4. The molar mass of the anhydrous salt is 136.1 g/mol; the molar mass of sulfate ion is 96.1 g/mol. 96.1 g is 55.8% of 172.2 g; this must be the molar mass of the hydrate. 172.2 – 136.1 = 36.1 g, which must be the mass of water in the hydrate. To the nearest whole number, this is two moles of water, so the formula for the hydrate is $CaSO_4{\cdot}2H_2O$.

Data and Calculations

Student data; will vary with nature of unknowns.

Answers to Postlaboratory Questions and Conclusions

1. a. Not all the sulfate ion would be precipitated. The percentage would be too low.
   b. The reported mass of product would be too high, making the percentage of sulfate too high.
   c. Since the product would appear to be lighter than it really is, the percentage of sulfate would be too low.

2. Loss of material in rinsing is the most likely source, followed by the dissolving of the barium sulfate precipitate if too large an excess of $BaCl_2$ is used.

3. $2H^+(aq) + CO_3^{2-}(aq) \rightarrow H_2O(l) + CO_2(g)$

4. $BaCO_3$ has low solubility in water; the precipitate could include traces of this salt, as well as $BaSO_4$, and the sulfate percentage would be too high.

5. Student first determines the molar mass of the anhydrous salt (the identity of the cation must be provided) and then follows the pattern used in Prelaboratory Question 4.

6. 0.117 g $BaSO_4$/233.4 g/mol $BaSO_4$ = 0.0005 mol of unknown. 0.071 g/0.0005 mol = 142 g/mol formula weight of unknown. The unknown is sodium sulfate, Na_2SO_4, formula weight 142 g/mol.

EXPERIMENT

12 Carbonate or Bicarbonate?

Placement: Stoichiometry

Approximate time for completion: 30 minutes

Approximate preparation time: 20 minutes

Materials: 24 of each unless noted otherwise, assuming a class of 24 students working individually

Apparatus

Milligram balance
Reaction tube (or 100-mm culture tube or test tube)
Electrically heated sand bath, adjusted to 30% or 40% power (no higher)
Plastic funnel
Boiling chips

Reagents

Amounts are for a class of 24 students working alone. Directions for preparation may be for more than the amount needed for one class; convenience has been taken into account.

Unknown solids, about 5 g (approximately 0.2 g per student)
Hydrochloric acid, 12 *M*

Discussion

This short experiment might well be paired with another, such as Experiment 13, "Hydrates: Structure and Properties." The student must deal with the concepts of limited reactants and stoichiometric ratios in analyzing the results, and decent technique is needed to correctly identify the "unknown."

The Prelaboratory Questions are intended to show students by example how experiments can be simplified through preliminary calculations.

Notes on the Procedure

1. In this case it is particularly important that students do the four Prelaboratory Questions in advance so that they know about how much of the acid they can expect to use.

2. The hydrochloric acid is best dispensed by a Beral-type pipette, which has been specifically designated for this purpose. Place one or two stock bottles of the concentrated acid in the hood or other central location. A test tube taped to the acid bottle makes a good holder for the dispensing pipette when it is not in use. In the prelaboratory instructions, remind students that neither the bottle stopper nor the pipette should be laid on the countertop. To reinforce this, provide a place for them to set the stopper (e.g., a small labeled beaker).

3. Place tubes containing approximately 0.2 g of solid near the balances so students will have a gauge when weighing out their samples.

4. *Substitutions:*
   a. 10-mL Erlenmeyer flasks may be used in place of the reaction tubes. If you make this substitution, have students use the same technique as was used in Experiment 8, "Synthesis of Manganese(II) Chloride," for drying moisture that condenses on the sidewalls.
   b. If you like, other unknowns may be used (e.g., K_2CO_3, $KHCO_3$, $CaCO_3$), but students must know what the cation is, and the chloride salts of the cation must be stable up to several hundred degrees Celsius.
   c. We usually recommend "boiling sticks" (wooden applicator sticks) for taming of boiling water and aqueous solutions. Here, their use is not indicated, since the solution can seep into the stick, affecting results.

Answers to Prelaboratory Questions

1. 0.200 g x (1 mol Na_2CO_3/106.0 g) = 1.89×10^{-3} mol

2. (38 g HCl/100 g solution) x (1.200 g solution/1.000 mL solution) x (1 mol HCl/36.5 g HCl) x (1000 mL/1L) = 12.5 *M*, or 13 *M* (2 significant figures)

3. 1.89×10^{-3} mol Na_2CO_3 x (2 mol HCl/1 mol Na_2CO_3) x (1000 mL HCl/12.5 mol HCl) = 0.302 mL

4. 0.200 g x (1 mol $NaHCO_3$/84.0 g) = 2.38×10^{-3} mol $NaHCO_3$; 2.38×10^{-3} mol $NaHCO_3$ x (1 mol HCl/1 mol $NaHCO_3$) x (1000 mL HCl/12.5 mol HCl) = 0.194 mL

Data and Calculations

Student data; will vary with unknowns and amounts used.

Conclusion

1. Student answers. From the mass of NaCl recovered, the number of moles may be calculated. The mass of the unknown is converted to mol $NaHCO_3$ and to mol Na_2CO_3. If the number moles of NaCl equals (within experimental error) the number of moles calculated for mol $NaHCO_3$, the identity is known. If the unknown was Na_2CO_3, the number of moles of NaCl recovered should be twice the number of moles calculated for the unknown as the carbonate.

Answer to Postlaboratory Question

1. The answer to the question should be, "Yes, based on the volume of hydrochloric acid needed: 0.302 mL (about 12 to 15 drops from a microtip pipette) for the carbonate versus 0.190 mL (9 to 10 drops) for sodium bicarbonate."

   Students, however, are more likely to try to seek a numerical answer such as the following: For sodium carbonate monohydrate, $Na_2CO_3{\cdot}H_2O$, the mass of NaCl produced would be 0.200 g $Na_2CO_3{\cdot}H_2O$ x (1 mol/124 g $Na_2CO_3{\cdot}H_2O$) x (2 mol NaCl/1 $Na_2CO_3{\cdot}H_2O$) x (58.5 g NaCl/1 mol NaCl) = 0.189 g, 3.26×10^{-3} mol, NaCl. Anticipated NaCl yield for Na_2CO_3 (anh.) is 2 x (1.89×10^{-3} mol) = 3.78×10^{-3} mol NaCl, and for $NaHCO_3$, 2.38×10^{-3} mol. Since the result for the monohydrate is closer to that expected for Na_2CO_3, the student would likely make the correct choice. Higher hydrates, however, would make the choice more difficult. Sal soda is listed as an *n*-hydrate.

EXPERIMENT

13 Hydrates: Structure and Properties

Placement: Stoichiometry

Approximate time for completion: 45 minutes

Approximate preparation time: 20 minutes

Materials: 24 of each unless noted otherwise, assuming a class of 24 students working individually

Apparatus

Milligram balance
Electrically heated sand bath, adjusted to 50% or 60% power
10 x 75 mm culture tubes, 42 (6 per student)
Glass stirring rod
1-mL graduated (if available) Beral-type pipette
Weighing paper cut into small squares (1.5 cm^2), 200 pieces
Watch glass, 15 cm diameter or larger
Wire test tube holders (or similar)

Reagents

Amounts are for a class of 24 students working alone. Directions for preparation may be for more than the amount needed for one class; convenience has been taken into account.

Part 1: 6 g each

Anhydrous cobalt(II) chloride, $CoCl_2$
Anhydrous calcium chloride, $CaCl_2$
Sodium carbonate decahydrate, $Na_2CO_3 \cdot 10\ H_2O$
Potassium aluminum sulfate (potassium alum), $KAl(SO_4)_2 \cdot H_2O$
Anhydrous copper sulfate, $CuSO_4$

Part 2: 2.5 g each

Calcium sulfate dihydrate
Copper(II) sulfate pentahydrate
Sucrose (table sugar)
Cobalt(II) chloride hexahydrate
Potassium chloride hydrate
Nickel(II) chloride hexahydrate
Potassium carbonate sesquihydrate ($1.5H_2O$)

Distilled water (in wash bottles)

Part 3:

Unknown: $CaSO_4 \cdot 2H_2O$; about 12 g (approximately 0.4 g per student)

Discussion

Although this experiment is easily completed in under an hour, a large amount of information is presented. Students experience the existence of and the difference between efflorescence and deliquescence, determine the percentage of water and the number of waters of hydration in a hydrate, and learn the distinction between carbohydrates and ordinary hydrated (ionic) solids.

The dehydration is carried out in triplicate to minimize random errors.

Notes on the Procedure

1. Note the need for unusually large watch glasses for Part 1. If 15-cm glasses are not available, consider using 2 or 3 smaller watch glasses or Petri dishes. Beakers work less well.

2. The crushing of the solids in Part 2 sometimes results in broken tubes. Caution students to use just enough force to break up the solids. Pyrex or Kimax tubes are preferable to soft glass for this purpose.

3. The heating in Part 2 can be done holding the tubes in the sand with the fingers, although we do not recommend doing so. As part of the prelaboratory discussion, emphasize that addition of water in Step 6 should wait until tubes cool and that addition is to be dropwise at first. Water can be removed from tube walls by pieces of rolled-up paper towel or filter paper.

4. *Substitutions:*
   a. Erlenmeyer flasks (10-mL) may be used in place of the reaction tubes in Part 3. If you make this substitution, have students use the same technique as was used in Experiment 8, "Synthesis of Manganese(II) Chloride," for drying moisture that condenses on the sidewalls.

b. Alternate unknowns could include any from the list of Part 2, plus copper(II) chloride dihydrate or even manganese(II) chloride tetrahydrate left from Experiment 8. In the case of substitutions, students must know either the identity of the cation or the molar mass of the anhydrous solid (as was given for calcium sulfate in the "Procedure.")

Answer to Prelaboratory Question

1. The mass (hence number of moles) of water is calculated from the amount of mass lost during heating. If the heating process is incomplete, the number of waters of hydration calculated will be too low.

Data and Calculations

Part 1: Student results: $CoCl_2$ goes from blue to pink; $CaCl_2$ particles swell and become less well defined. Sodium carbonate and potassium alum will show little change in appearance, but some water is lost through deliquescence; this appears as a wetness on the paper. As it hydrates, copper(II) sulfate goes from pale bluish-white to light blue; the color may be irregular.

Part 2
Calcium sulfate: white; yes; white; yes; yes
Copper sulfate: blue; yes; bluish-white; yes; yes
Sucrose: white; yes; brown; no; no
Cobalt chloride: pink; yes; blue; yes; yes
Potassium chloride: white; no; white; yes; no
Nickel chloride: green; yes; yellow-green; yes; yes
Potassium carbonate: white or clear (slight); yes; white; yes; yes

Part 3: Student data; agreement among trials is usually quite good.

EXPERIMENT

14 Periodic Trends in Physical Properties: Densities of Unknowns

Placement: Mendeleev, periodicity, and the periodic law

Approximate time for completion: 45 minutes

Approximate preparation time: 20 minutes

Materials: 24 of each unless noted otherwise, assuming a class of 24 students working individually

Apparatus

Milligram balances
Graduated cylinder, 10-mL (Part 1)
Graduated cylinder, 25-mL (or 50-mL) (Part 2) (not listed in the laboratory manual)

Reagents

Amounts are for a class of 24 students working alone. Directions for preparation may be for more than the amount needed for one class; convenience has been taken into account.

Part 1

Granular silicon, 125 g
Tin shot, 125 g
Water (tap water will work in most locations)

Part 2

Pennies, 600 (25 per student)

Discussion

The first part of the experiment is intended to give students a feel for the meaning of the periodic law. The ability to predict the densities of some previously unknown elements was one of the major triumphs of the law, as stated by Mendeleev (and Lothar Meyer). The procedure is both simple and familiar and so should present no challenge to students. In fact, it has been our experience that good results are so easily obtained that students can focus clearly on their significance rather than on attaining them.

The second part is intended simply to let students see how the same property can be measured in different ways and to help them learn to evaluate the different methods in terms of accuracy and precision.

We have not specified whether the pennies are to be pre-1982, post-1983, or a combination despite the fact that the composition of the pennies changed radically during the years 1982–1983. Over the past two years using this experiment, one of us (JGL) has found that the distribution in a random assortment is so predominantly loaded with the newer pennies that student results are very much in accord with the post-1983 composition. When the average density of the pennies is compared with the densities of zinc and copper as given in Table 14.1, it is clear that (1) there is very little variation from one student to the next and (2) the average density closely approximates that of zinc. This may come as a great surprise to many students.

Notes on the Procedure

1. If more than one section is to use the same solids (Sn, Si, pennies), it will be necessary to find a way to dry them between uses. A quick rinse in a beaker of acetone followed by spreading on absorbent paper towels in a well-ventilated area usually accomplishes drying within an hour or two.

2. Small plastic funnels or funnels made from filter paper or rolled paper sheets are helpful in getting the pieces of silicon and tin into the graduated cylinder.

3. There is nothing magical about the choice of 25 pennies; 15 or more will do. We were after a number that was large enough to average out random variations.

Answer to Prelaboratory Question

1. $(3.51 + 1.54) \div 2 = 2.525 = 2.53$ g/cm^3. Deviation = $2.6 - 2.5 = 0.1$ g/cm^3; percentage error = $(0.1 \div 2.6) \times 100\% = 3.85 = 4\%$ error. Students almost always report too many significant figures; this is a good reminder.

Data

Student results.

Calculations and Conclusions

1. Student results; expect values close to the accepted values unless students forget to subtract the mass of graduate and water. Accepted values: Si = 2.33 g/cm^3; Sn = 6.52 to 6.56 g/cm^3. [This is an average for the ordinary beta (β) or "white" form and the alpha (α) or "gray" form, which exists at temperatures below 13.2°C. A third allotrope, γ-Sn, exists above 161°C. *Source:* Kleinberg et al., *Inorganic Chemistry,* D. C. Heath, Lexington, Mass., 1960.]

2. Using the literature values above (Sn = 6.54), the average is 4.44 g/cm^3. Student results should be in fairly good agreement with this.

3. Reported for Ge: 5.35 g/cm^3. Students will be close to the expected value of 4.44 and so will have fairly high percentage errors. The percentage error for 4.44 is 17.0%, for example. Expect a lot of "What did I do wrong?" questions.

4. a. Student answers will vary somewhat. Typical result: h = 3.65 cm; r = 1.85 cm; $V = \pi r^2 h$ = 39.2 cm^3. Many students will round to the nearest 0.1 cm on h and r, which in this example would produce a volume of 41.96 or 42 cm^3.
   b. Unpredictable, but our experience has been that volumes measured by water displacement tend to be within about 1 cm^3 of the calculated value. Deviations are random.
   c. Again, deviations are random, but generally ±1 cm^3.
   d. Answers vary, but expect values in the range of 7.4 to 7.7 g/cm^3.

5. a. Depending on the size and style of graduate used in Part 2, it is likely not possible for students to estimate volume to the nearest tenth of a cubic centimeter, while three significant figures are to be expected from the "ruler" method. To this extent, the latter is more precise. If equal skill levels are brought to both methods, there would be little to choose for accuracy, but students seem to have more confidence in the water displacement, notwithstanding the errors possible due to splashing and estimating of meniscus levels.
   b. Answers will vary with student data. Students take the difference between their individual answers to 4d and 4e, divide that difference by whichever of 4d and 4e they selected in 52, then convert the resulting decimal fraction to a percent.

6. Most, if not all, students will know to focus on the densities of copper and zinc. It should be very evident that the average density of the pennies reflects the fact that most are almost entirely zinc, with only a thin layer of copper veneer.

Answers to Postlaboratory Questions

1. So long as there is no chemical reaction between the liquid and the pennies, and assuming that the liquid has a density of less than about 7 g/cm^3, all liquids should give the same result. Nitric acid and mercury are among the liquids eliminated.

2. You would expect to float.

3. Assuming that the ice is floating, the water level will rise 3 mL. After it melts, the water level will not rise any more. If the ice is submerged, the level of the water will rise 3.33 mL, and after it melts, the level will drop by 0.33 mL.

EXPERIMENT

15 Distillation

Placement: Physical properties of liquids, industrial chemistry

Approximate time for completion: 2 hours

Approximate preparation time: 30 minutes

Materials: 24 of each unless noted otherwise, assuming a class of 24 students working individually

Apparatus

Sand bath
Distilling apparatus
    5-mL long-necked round-bottomed flask
    Connector with support rod
    Distillation head/condenser
    Thermometer adapter
1-mL graduated pipette
Thermometer, mercury or digital
Vials, 2
30-mL beaker
Three-prong clamps
Clamp holders
Ring stand
Metal spatula
Copper sponge (Chore Boy), 1 g per student

Reagents

Amounts are for a class of 24 students working alone.

Part 1

Methyl alcohol, 50 mL
Water, 50 mL
Boiling chips
Copper sulfate or, better, methylene blue dye or blue food coloring

Part 2: One of the following 50:50 binary mixtures, 4 mL per student

Acetone (bp 56.5°C) and 2-propanol (bp 82.4°C)
2-Propanol (bp 82.4°C) and water (bp 100°C)
2-Propanol (bp 82.4°C) and 1-butanol (bp 117.2°C)
Acetone (bp 56.5°C) and water (bp 100°C)
Water (bp 100°C) and 1-butanol (bp 117.2°C)

Discussion

The object of this experiment is to illustrate the usefulness of distillation as a means for separating and purifying substances. No attempt has been made at this early stage of the course to go into such matters as phase diagrams, Raoult's law, vapor pressure versus temperature, azeotropes, and so on.

The first experiment shows that relatively pure methanol can be distilled from a 50:50 mixture of methanol and water. Proof that a relatively pure organic compound has been obtained is provided by the fact that it can be set afire in the hood.

It is very instructive to have each student add to the mixture a ***minute*** quantity of methylene blue or similar intense dye in order to demonstrate that the nonvolatile organic dye does not distill. If difficulty is encountered in having students add too much dye, then the methanol or the 50:50 mixture of methanol and water can be dyed before dispensing to the students. A crystal of copper sulfate, dissolved first in the water before the methanol is added, also illustrates the same effect.

In the second experiment five binary mixtures are provided. Each student is to identify the components from the boiling points. The apparatus should be reasonably dry before adding the mixture. A much better separation can be achieved by packing the long neck of the flask with about 0.75 g of copper sponge (Chore Boy), which must be carefully dried between experiments. The extraordinary slowness of the distillation (one drop every 30 seconds) cannot be overemphasized. Only at this very slow rate does one achieve the clean separation of mixtures that this apparatus is capable of.

Notes on the Procedure

1. Water and isopropyl alcohol form a binary azeotrope that boils near the boiling point of the alcohol (80.3°C). Since it contains a high (87) percentage of the alcohol, it is usually possible to get enough of the aqueous layer to identify the components, but confidence is shaken.

2. Even though students have used the sand baths several times by this point, it is important to show the placement of the distilling flask just barely in contact with the sand. This is to make it easy to adjust the heating without resetting the thermostat.

3. If you are interfacing the experiment with computers and are not using the software named in the manual, provide students with instructions on collecting the data.

Answers to Prelaboratory Questions

1. Some commercial items that are prepared or purified by distillation include antifreeze, turpentine and all other paint solvents, and fingernail polish remover.

2. The presence of water and/or methanol would contaminate the mixture used for Part 2. Acetone's high volatility ensures much more rapid drying than would water.

Data and Observations

Part 1: Student results; it is not uncommon for less than half the drops to boil at the temperature characteristic of methanol. This results from too-rapid heating. The color (if used) will remain with the pot residue. The distillate is flammable, but the 50:50 mixture is not.

Part 2: Except in cases of very poor technique or rushing of the process, identification is nearly always correctly done. You can help by being careful to pair one high-boiling liquid with one of the lower-boiling ones.

Answers to Postlaboratory Questions

1. a. The boiling point will be raised.
   b. There will be no effect on the boiling point.
   c. 100°C

2. Additional heat must be supplied (the heat of vaporization) in order for the water to pass into the vapor state.

3. 18 g of water (one mole), for example, when vaporized, will occupy 22.4 L. If this is not completely condensed, there will be a very large pressure increase in a closed apparatus.

4. Reduction of the pressure will lower the boiling point. Methanol can be separated from water at reduced pressure, but the boiling points will be somewhat closer to each other.

5. Yes. By reducing the pressure over the water sufficiently, water can be made to boil below room temperature. This is often the basis of a lecture demonstration that is carried out as follows: A 60-mL plastic syringe fitted with a Leur stopcock is half filled with water at 50 to 60°C. The stopcock is closed. When the plunger is withdrawn, the water will be seen to boil.

EXPERIMENT

16 Avogadro's Law

Placement: Gas laws, ideal gas law, Avogadro's law

Approximate time for completion: 2 hours

Approximate preparation time: 30 minutes

Materials: 24 of each unless noted otherwise, assuming a class of 24 students working individually

Apparatus

Milligram balances
Bottle and syringe apparatus (see Figure 16.1 in the student text)

Reagents

Unless noted otherwise, amounts are for a class of 24 students working individually.

Sodium bicarbonate, 25 g
White vinegar, 360 mL
Glycerol (for lubrication of syringe)

Notes on the Procedure

1. As the Prelaboratory Questions and answers point out, the solubility of carbon dioxide in water (dilute carbonic acid) is appreciable and probably makes a major contribution to the error in calculation of the molar volume. One way around this problem, although

more hazardous, would be to use a small volume of a more concentrated acid. A single milliliter of concentrated hydrochloric acid (12 mmol) would suffice for five successive experiments. After the first experiment, the liquid would be saturated in carbon dioxide. The problem with this scheme is ascertaining that all the bicarbonate has reacted with the small volume of acid.

2. Another possible source of error is due to friction in the syringe. The volume determined experimentally will be slightly smaller than it would be at 1.0 atm pressure because of this friction. If some means could be devised to measure the pressure of the gas in the syringe, this error could be minimized by adjusting the syringe until the internal pressure is 1 atm. The use of a T-connector between the bottle and the syringe connected to a pressure-measuring device such as an open-end manometer or a Vernier pressure gauge would allow the pressure within the syringe to be adjusted to 1 atm.

Answers to Prelaboratory Questions

1. 0.06 L/22.4 mol L^{-1} = 2.68 x 10^{-3} mol of carbon dioxide is produced, which is equal to (84.01 g/mol)(2.68 x 10^{-3} mol) = 0.225 g of sodium bicarbonate.

2. The *CRC Handbook of Chemistry and Physics* lists the solubility of carbon dioxide as 0.145 g per 100 cm^3 H_2O at 25°C. 0.145 g CO_2 ÷ 44.0 g/mol = 3.30 x 10^{-3} mol in 100 mL, or 3.30 x 10^{-2} mol/L.

3. a. 3.30 x 10^{-3} mol in 100 mL corresponds to 3.30 x 10^{-4} mol in 10.0 mL.
   b. V = 3.30 x 10^{-4} mol x 0.0821 atm·L/mol·K x 298 K/1.00 atm = 8.06 x 10^{-3} L = 8.06 mL

Data Table

Mass of $NaHCO_3$ (g)	Volume of CO_2 (mL)	Mass of $NaHCO_3$ (g)	Volume of CO_2 (mL)
0.160 g	37.5 mL	0.222 g	54.5 mL
0.174 g	45.5 mL	0.226 g	55.0 mL
0.201 g	45.0 mL	—	—
0.214 g	48.0 mL	—	—

Calculations

1. 1—0.00190; 2—0.00207; 3—0.00239; 4—0.00255; 5—0.00264; 6—0.00269

2. 1—19.7; 2—22.0; 3—18.8; 4—18.8; 5—20.6; 6—20.4 L/mol.
   Avg = 20.1 ± 0.5 L/mol

Conclusions

The 8-mL difference calculated in Prelaboratory Question 3 will have a major effect on the molar volumes. Addition of 8.1 mL to each of the sample volumes shown in the Data Table would give the following molar volumes for the six trials: 1—23.9; 2—25.9; 3—22.2; 4—22.0; 5—23.7; 6—23.5 L/mol. The average molar volume is now 23.5 ± 0.8 L/mol.

EXPERIMENT

17 Charles' Law: Temperature/Volume Relationships for Gases

Placement: Gas laws, ideal gas law

Approximate time for completion: 2 hours

Approximate preparation time: 30 minutes

Materials: 24 of each unless noted otherwise, assuming a class of 24 students working individually

Apparatus

Thermometer
Sealed melting-point capillary
Small rubber bands cut from gum rubber tubing
Beakers with water at five different temperatures: near zero (ice), room temperature, 40, 60, 80, and close to 100°C.

To prepare the apparatus, a small drop of silicone, cottonseed, or mineral oil is introduced about 2 cm from the open end of the melting-point capillary using a Pasteur pipette drawn to a fine capillary.

Notes on the Procedure

Because they are safer, have a faster response time, and are more accurate, we recommend the use of digital thermometers for this experiment. It may take some searching to find mercury or alcohol thermometers that have graduations near the bulb. Ordinary 3-in. immersion thermometers will necessitate the use of very large beakers to give sufficient water depth.

Answer to Prelaboratory Question

1. No. The diameter of the capillary does not affect the results. A larger-diameter capillary will show a smaller change in length of the air column than a small-diameter capillary of the same volume.

Conclusion

1. Lack of ability to measure the length of the air column precisely, the fact that the air column is not a perfect cylinder (it is rounded at both ends), the fact that air, especially humid air, is not an ideal gas, the fact that the range of temperatures is not terribly large, and the fact that the results must be extrapolated almost 300°.

2. As the air cools, the molecules of nitrogen and oxygen have less kinetic energy to expend pushing against the oil droplet at one end of the capillary.

3. The primary assumption is that the individual gas molecules occupy a very small percentage of the total volume and thus do not interact with each other to any large extent.

EXPERIMENT

18 Determination of the Molar Volume of a Gas and the Universal Gas Constant

Placement: Gas laws

Approximate time for completion: 45 minutes

Approximate preparation time: 1 hour

Materials: 24 of each unless noted otherwise, assuming a class of 24 students working individually

Apparatus

Vernier calipers or rulers (4 to 6, at dispensing station, especially if calipers are used) (Note that there is a good ruler on page 365 of the text.)
Graduated cylinder, 10.0 mL
Rubber stopper, 1-hole, 00 (See "Notes on the Procedure")
Copper wire, 18 to 22 gauge, 300 cm
Barometer (one for class)
Thermometer
Wash bottle of distilled water

Reagents

Amounts given are more than enough for a class of 24 students working individually. See "Notes on the Procedure."

Magnesium ribbon, 75 cm, cut into 0.9- to 1.0-cm lengths
Hydrochloric acid, $HCl(aq)$, 3*M* *(Slowly and with constant stirring, add 50 mL of concentrated acid to 150 mL of distilled water. Allow the solution to adjust to room temperature before using. Corrosive fumes are generated; work in a fume hood and wear protective clothing.)*
Sodium bicarbonate, $NaHCO_3$ (for neutralizing residual acid after each trial), 75 g

Discussion

The experiment is scaled down from an older version that may have first been reported by Michael Faraday. Assuming that students are careful and that they determine the length of magnesium pieces ± 1 mm, two significant figures are possible, and results rival those obtained with 50-mL gas-measuring tubes (eudiometers). Disposal is minimal on this scale, and three trials may be done easily in under an hour. The calculations take time.

Values for R generally run slightly above the accepted value of 8.31 kPa·L/mol·K.

Notes on the Procedure

1. If you can supply calipers, it may be worthwhile, if only because students become more aware of non-right-angle cuts in the strips. An inexpensive metric Vernier caliper can be obtained at a hardware store or machinist's supply store. A plastic one works well and will not rust. With this device, the length of the ribbon can be measured to 0.01 mm. This device, as well as the barometer, gives students practice in reading a Vernier. They should estimate the length to the nearest 0.01 to 0.02 mm using the ruler on page 365 of the student text in order to avoid making serious errors in reading the Vernier.

   Values of the molar volume of hydrogen at STP are routinely within the same range with this method as they are when 50-mL gas-measuring tubes are used. The micro version has the advantage that each student can complete three trials within a single laboratory period. It also removes the need for a separate, deep container for making the water levels equal.

2. *Magnesium ribbon.* The directions here assume a thickness of 0.15 mm (e.g., Flinn Scientific, M0001, Aldrich 26,630-2); thickness may vary by supplier. Thicker ribbons mean either a shorter length must be used or a larger graduated cylinder must be substituted.

   To minimize variations, clean about 1 m of the ribbon just before you cut and weigh the 100.0-cm strip. Use this sample to prepare samples for students.

   Best results are obtained when a little over 1 m of the ribbon is rubbed lightly with 0000 steel wool and then wiped clean just before weighing the 100.0-cm length. This magnesium is then used to make 0.90- to 1.00-cm lengths that the students used. The linear density thus obtained is slightly above 0.90 g per 100.0 cm and does not generate volumes of hydrogen above 10.0 mL. If the linear density is below 0.9 g per 100 cm, then lengths up to 1.05 cm will probably be fine, and another significant figure is possible.

   Washing the apparatus with dishwashing detergent and a drop of detergent in the beaker will facilitate clean drainage of the graduated cylinder.

   A systematic correction of 0.2 mL on all volume readings has been suggested (G. S. Levinson, *J. Chem. Educ.* 59: 337, 1982) for this experiment because the meniscus is read upside down. It has been our experience that the values obtained using Pyrex graduates (3075) with red plastic hex bases (which are removed for this experiment) do not benefit from the correction.

   To save prep time, cut the ribbon in to 3- to 4-cm lengths yourself and then have students make the final division into appropriate lengths.

3. In cases where the barometric pressure is low, molar volume will be higher, making it necessary to use lengths less than 1 cm. You may have to let students discover this by trial and error. If one trial goes off-scale, students can be challenged to seek a strategy for determining the actual volume of gas generated. (This is an excellent question for a prelaboratory quiz!) One possibility is to mark (wax pencil) the water level in the graduate and then later determine the volume of water needed to fill the upright graduate to that line. Another option is to trap the water that remains in the graduate with a solid stopper and then set the graduate upright on the bench top. Assuming the graduations extend low enough on the sides, the student can then read directly the amount by which the volume exceeded 10.0 mL.

4. One or two boxes of ordinary baking soda will suffice for the neutralization during cleanup.

Answers to Prelaboratory Questions

1. If the pressure of the hydrogen were assumed equal to the atmospheric pressure, the number of moles in the graduated cylinder would come out too high, making the molar volume, and *R*, too small.
   a. The gas in the cylinder will be a mixture of two gases, hydrogen and water vapor. If it is assumed, incorrectly, that this is all hydrogen, then the molar volume calculated for hydrogen will be too large.
   b. Yes. If the volume of gas calculated is too large, then *R* also will be too large. $R = (PV/nT)$.

2. 3×10^{-3} L x (3 mol HCl/L) x (1 mol $NaHCO_3$/1 mol HCl) x 84 g $NaHCO_3$/mol = 0.76 g $NaHCO_3$ = 0.8 g $NaHCO_3$ (about 1 g is good enough.)

Answers to Postlaboratory Questions

1. Calculation 7. Molar volume is expressed in liters per mole.

2. Yes; students almost invariably find evidence that molar volume is constant, whether they consider only their own results or those of an entire laboratory section.

3. $V = nRT/P = 1.00 \times R \times 273$ K/101.3 kPa.

4. [(difference from 22.43) ÷ (22.43)] x 100% = % error.

5. Students will recognize the imprecisions of measuring the magnesium ribbon. Most will see, too, that oxidation of the metal can affect results. A few understand that we are assuming, perhaps incorrectly, a uniform thickness for the ribbon.

6. The other product of reaction, magnesium ions and chloride ions, remains in solution.

EXPERIMENT

19 Oxidation-Reduction Reactions

Placement: When oxidation and reduction are first introduced

Approximate time for completion: 1 hour

Approximate preparation time: 1.5 hours

Materials: 24 of each unless noted otherwise, assuming a class of 24 students working individually

Apparatus

96-well test plate
Hand lens (optional)
Tweezers or forceps for handling metal pieces
Safety goggles

Reagents

Except as noted, directions are given for preparing 50 mL of solution. A single class of 24 students working alone will need far less than is prepared.

Copper and zinc pieces. Cut bits about 3-mm square from sheet stock; pieces should be large enough to see but small enough for the test plate wells. If the zinc has an oil coating, rinse it with acetone before cutting.

Magnesium pieces. Cut pieces of ribbon equal in length to the ribbon width; clean oxidized ribbon with steel wool (0000) before cutting.

1.0 *M* hydrochloric acid, HCl (*Slowly and with constant stirring,* add 4 mL of concentrated acid to 46 mL of distilled water. Allow to cool before using.)

1.0 *M* copper(II) nitrate (6.04 g $Cu(NO_3)_2 \cdot 3H_2O$ is dissolved in 25.0 mL of distilled water. A less-concentrated solution does not give sufficiently detectable results.)

1.0 *M* zinc nitrate (7.34 g $Zn(NO_3)_2{\cdot}6H_2O$ is dissolved in 25.0 mL of distilled water)
0.1 *M* iron(III) chloride (1.35 g $FeCl_3{\cdot}6H_2O$ to make 50 mL with 0.01 *M* HCl; prepare the solution with distilled water, and then add 2 to 3 drops of concentrated HCl.)
0.1 *M* iron(II) sulfate (1.39 g $FeSO_4{\cdot}7H_2O$ with 2 to 3 drops of 18 *M* H_2SO_4 added in water to make 50 mL.
0.1 *M* potassium permanganate (0.79 g $KMnO_4$ in water to make 50 mL)
Potassium iodate, 0.1 *M* (1.07 g KIO_3 in water to make 50 mL)
Potassium iodide, 0.5 *M* (4.15 g KI; make fresh, because shelf life is poor.)
3 *M* sulfuric acid, H_2SO_4 (*Slowly and with constant stirring*, add 17 mL of concentrated acid to 83 mL of distilled water. This solution is used again in other experiments; see Experiment 20, "Analysis by Oxidation-Reduction Titration." It might be a good idea to prepare more than 100 mL.) *Caution:* Large quantities of heat are evolved; allow to cool before using.
0.1 *M* ammonium vanadate, 100 mL (0.234 g NH_4VO_3 in 50 mL 1.0 *M* H_2SO_4. The solid will dissolve slowly, eventually giving a bright yellow color similar to chromate.)
0.1 *M* vanadyl sulfate (0.5 g Na_2SO_3 is added to 50 mL of the ammonium vanadate prepared above. Boil 5 minutes to remove excess sulfite as SO_2. Do this in front of your students during prelaboratory preparation; a white powder is added to a bright yellow solution; the mixture goes through green and blue.
0.1 *M* sodium sulfate (0.71 g Na_2SO_4 in water to make 50 mL)
0.1 *M* copper(II) sulfate (1.25 g $CuSO_4{\cdot}5H_2O$ in water to make 50 mL)
3% hydrogen peroxide, commercial disinfectant
Chlorine water [In a fume hood, slowly add 5 mL of 3 *M* H_2SO_4 (above) to 20 mL of household bleach, NaOCl(*aq*). Dilute to 100 mL; store in a tightly sealed bottle. Use glass; chlorine will attack plastic.]
5% sodium hypochlorite, household liquid bleach (Clorox, unscented)

Discussion

While this is a lengthy experiment to set up and very reagent-intensive, it has been designed to cover the full range of redox systems, from single replacement to disproportionation. The questions that accompany the various parts of the experiment emphasize terminology and deductive logic.

Notes on the Procedure

1. Cleaning up is perhaps best accomplished by shaking the inverted test plate over a large tray lined with several thicknesses of paper towel. Stray bits of metal may have to be removed with the tweezers. Transition metal ion solutions will evaporate in a few days, leaving solid waste, which is generally easier and less expensive to dispose of.

2. *What students will see:*

   Part I: The magnesium will dissolve with evolution of (hydrogen) gas.

Part II: The zinc will appear to darken in the copper solution as copper metal forms on its surface. The blue color begins to fade, but this change is not easily detected. If students have trouble, suggest that they put some of the copper solution in an adjacent well for comparison. No reaction occurs when copper is added to the zinc solution. Students may not be happy with this, but they need to accept that many of these reactions are essentially irreversible; hence, if zinc metal reactions with Cu^{2+} ions, copper metal will not react with zinc ions. This is also a good time to revisit the activity series for metals (Experiment 27).

Part III: Iron(II) reacts; iron(III) does not. The purple color changes but may stop at the red-brown of MnO_2 rather than Mn^{2+} if the system is not acidic enough. Students who are not careful with the dropwise addition of permanganate may not recognize the fact that the first drop or two lost the purple color; they certainly will not be able to determine the relative numbers of drops needed. These students might benefit from repeating the step. The total number of drops should be about 10, since each mole of Fe^{2+} loses 1e, while MnO_4^- is gaining 5e per mole. If the reaction stops at MnO_2, then 6 drops total should be required.

Part IV: A dark red color characteristic of iodine results. The change is quite dramatic, but caution students that the product is toxic. Some people have allergic reactions to iodine on their skin. Disposal may require special steps in some jurisdictions, although very little is involved here.

Part V: VO_2^+ (from the vanadate: $VO_3^- + 2\ H^+ \rightarrow H_2O + VO_2^+$ is yellow; VO^{2+} (vanadyl ion) is blue. Encourage students to refer to them as vanadium(V) and vanadium(IV), respectively; it will save much confusion. The correct name for the yellow VO_2^+ is dioxovanadium(V).

The reduction potentials follow:

$MnO_4^- + 8H^+ + 5e \rightarrow Mn^{2+} + 4H_2O$	(1.51*v*)
(Bleach) $HClO + H^+ + 2e \rightarrow Cl^- + H_2O$	(1.48*v*)
$Fe^{3+} + 1e \rightarrow Fe^{2+}$	(0.77*v*)
$SO_4^{2-} + 4H^+ + 2e \rightarrow H_2SO_3 + H_2O$	(0.17*v*)
$H_2O_2 + 2H^+ + 2e \rightarrow 2H_2O$	(1.78*v*)
$Cu^{2+} + 2e \rightarrow Cu$	(0.34*v*)
$VO_2^+ + 2H^+ + e \rightarrow VO^{2+} + H_2O$	(0.99*v*)

Answers to Prelaboratory Questions

1. The oxidizing agent takes electrons from the substance being oxidized; the reducing agent supplies electrons to the substance being reduced. The oxidizing agent gets reduced; the reducing agent gets oxidized.

2. In disproportionation reactions, the same species is both oxidized and reduced.

3. There are more from the transition metals than from any other portion of the table.

Conclusions

Part I

1. a. Mg^0
   b. Mg^{2+}
   c. Oxidized; electrons were lost.
   d. $Mg \rightarrow Mg^{2+} + 2e$

2. $2H^+ + 2e \rightarrow H_2$

3. $Mg + 2H^+ \rightarrow Mg^{2+} + H_2$

Part II

4. a. Zinc in copper(II) nitrate darkened; copper in zinc nitrate had no reaction.
   b. $Cu^{2+} + Zn \rightarrow Cu + Zn^{2+}$ is the correct choice; it shows metallic zinc reacting with Cu^{2+} ions.
   c. Cu^{2+} is reduced and is the oxidizing agent; Zn is oxidized and is the reducing agent.

Part III

5. Fe^{2+} decolorized the permanganate.

6. They do not react in a 1:1 ratio; the astute student may deduce the correct 5:1 ratio.

7. a. $Fe^{2+} \rightarrow Fe^{3+} + e^-$; iron(II) was the reacting species.
   b. Oxidized; it loses $1e^-$.
   c. Reduced; since it must be the oxidizing agent, it is being reduced.
   d. The oxidizing agent is permanganate; the reducing agent is iron(II).

Part IV

8. Oxidation: $2I^- \rightarrow I_2 + 2e$; reduction: $2IO_3^- + 12H^+ + 10e \rightarrow I_2 + 6H_2O$.

Part V

9. MnO_4^-, H_2O_2, OCl^-, IO_3^-, Cl_2; iodate's effect may be difficult to detect.

10. a. MnO_4^-, H_2O_2, OCl^-, IO_3^-, Cl_2, and VO_2^+; again, the effect of iodate may be difficult to detect.
    b. To complete the comparison between Fe^{2+}/Fe^{3+} and VO^{2+}/VO_2^+.
    c. Iron(II) is more easily oxidized because VO_2^+ can oxidize Fe^{2+} but Fe^{3+} cannot oxidize the VO^{2+}.

EXPERIMENT

20 Analysis by Oxidation-Reduction Titration

Placement: After Experiment 19, during discussions of redox systems. If this and Experiment 21, "Concentration of Hydrogen Peroxide," are both to be done, assign Experiment 21 first; it is less sophisticated and will build technique and confidence.

Approximate time for completion: 1.5 hours if students prepare their own iron(II) unknown solution.

Approximate preparation time: 30 minutes

Apparatus

Milligram balances (at least one for every 6 students; 1 for 4 is better)
Microtip pipettes
Label material (see Note 3 of "Notes on the Procedure")
10-mL Erlenmeyer flasks (3 per student, if possible)
Small hot plates, sand baths, or steam baths (see Note 4 of "Notes on the Procedure")
10- or 25-mL volumetric flasks (see Note 1 of "Notes on the Procedure")

Reagents

Except as noted, directions are given for preparing 50 mL of solution. A single class of 24 students working alone may need far less than is prepared.

3.0 *M* sulfuric acid, H_2SO_4. *Slowly and with constant stirring,* add 200 mL of concentrated acid to 1.0 L of distilled water. *Caution:* Large quantities of heat are generated; allow to cool before using. The solution is used as the solvent for the iron unknown in Part 2, so expect to need about 700 mL per section.

Potassium permanganate, $KMnO_4$. Place 0.27 to 0.3 g (±1 mg) of solid in a tared beaker. Add 100 mL of distilled water and stir to dissolve. This gives a concentration of about

2.2 x 10^{-5} mol $KMnO_4$/g of solution. Students will standardize the solution, so you need not reweigh beaker and contents.

Sodium oxalate, $Na_2C_2O_4$. Between 0.57 and 0.60 (±1 mg) of solid that has been dried at 105 to 110°C for several hours and then cooled in a desiccator is placed in a tared beaker. Add 100 mL of distilled water; reweigh. The solution will have a concentration of about 4.3 x 10^{-5} mol $Na_2C_2O_4$/g. Students must be given the exact concentration to three significant digits.

Unknown iron salt. See Note 4 of "Notes on the Procedure."

Concentrated phosphoric acid, H_3PO_4. One small dropper bottle will do; only about 3 mL is needed for an entire class.

Discussion

This titration experiment begins with standardization of the titrant for use in later analysis. The system is a familiar one and can be found (on a larger scale) in many standard manuals. The glassware and techniques of quantitative analysis are introduced as students prepare their own unknown solutions. Use of small-scale apparatus cuts the time roughly in half compared with macro versions.

An alternative to the goal of calculating the mass percent of iron in the unknown is to tell students what iron salt (sulfate, ammonium sulfate, etc.) they have and then ask them to determine the numbers of waters of hydration. A third possibility, but one that will take significantly longer, is to have students dissolve a weighted piece of iron alloy (coin, wire, razor blade) and then divide the solution (See Experiment 7, "Isolation of the Silver in a Dime") and analyze their own aliquot.

Notes on the Procedure

1. Volumetric flasks were inadvertently omitted from the Materials list. The procedure specifies 25-mL flasks because we assumed that they would be more likely to be already in stock. The 10-mL size reduces waste.

2. a. As noted in the procedure, the oxalate solution is warm enough to titrate when steam has begun to mist the flask walls. Remind students that they do not have to warm the flasks for the permanganate-iron(II) titrations.
   b. One of the products of the oxalate titration in Part 1 is carbon dioxide, which can undergo hydrolysis to produce the weak acid H_2CO_3. It would seem that this should ensure that there would not be a lack of hydrogen ion, but this does not seem to be the case. In fact, students frequently encounter problems of MnO_2 formation, perhaps due to the elevated temperature.

3. The best labeling tape we have found is available from many hardware and electrical supply stores: 3M's Scotch 35 electrical tape. To find the local suppliers, call Scotch Electrical Products at 1-800-245-3573. they will give you the names of local distributors. We recommend that you contact one of these and ask them to order a case of the tape for you (12 boxes, 10 rolls each, about $130). This will last 2 or 3 years. It

is available in a number of colors, but the yellow tape (Part No. 10257) makes the most distinctive labels.

Ultra-fine tip Sharpies make excellent, fast-drying labels.

4. If you have the budget, consider ordering the small Thermoplate hot plates from Laboratory Craftsmen, especially if you already have the controllers for the Thermowell sand baths. These are small and very reliable and heat a 259-mL beaker of water to boiling in about 5 minutes.

5. Iron(II) sulfate heptahydrate or ammonium iron(II) sulfate hexahydrate both make good unknowns. One of us (JGL) uses the latter, to which small, measured amounts of sodium sulfate have been added to provide a variety of unknowns.

Answers to Prelaboratory Questions

1. $8H^+(aq) + 5Fe^{2+}(aq) + MnO_4^-(aq) \rightarrow 5Fe^{3+}(aq) + Mn^{2+}(aq) + 4H_2O(l)$

2. a. $16H^+ + 5C_2O_4^{2-} + 2MnO_4^- \rightarrow 10CO_2 + 2Mn^{2+} + 8H_2O$
   b. Oxidizing agent is permanganate; oxalate is the reducing agent.
   c. 2:5

Answers to Postlaboratory Questions

1. Less permanganate would be consumed, so less iron would be reported. Too low.

2. The oxidation state of Mn is changing only by 3, not 5, so for part of the system the mol ratio (MnO_4^-:Fe) is changed from 1:5 to 1:3. Assuming that the calculations are done using the 1:5 ratio, the value for the amount of iron present will be low, but the extent of the effect cannot be predicted.

3. In the second case, drops are hard to control, and the conversion to milliliters would introduce significant uncertainty. If burettes are used, the first difficulty is that the concentration units are designed for gravimetric not volumetric titration. In addition, stains to burettes are a real problem.

4. a. For $FeSO_4 \cdot 7H_2O$, 20.1% Fe; for $Fe(NH_4)_2(SO_4)_2 \cdot 12H_2O$, 11.2% Fe. As noted earlier, the compound is actually a 6-hydrate, and the mass percentage of iron is 14.2%.
   b. The mass percentages are 23.1% for the pentahydrate and 20.1% for the heptahydrate. A student with good technique for should be able to make the distinction easily, since they differ by significantly more than the expected experimental uncertainty.
   c. Temperatures near the boiling point of water should be high enough to decompose the compound by loss of ammonia. The *CRC Handbook of Chemistry and Physics* reports decomposition at 100 to 110°C.

EXPERIMENT

21 Concentration of Hydrogen Peroxide

Placement: Oxidation-reduction or with discussion of titrations. If this experiment and Experiment 20, "Analysis by Oxidation-Reduction Titration," are both to be done, this one should be done first.

Approximate time for completion: 40 minutes

Approximate preparation time: 30 minutes

Materials: 24 of each, unless noted otherwise, assuming a class of 24 students working individually

Apparatus

10-mL Erlenmeyer flasks, 72 each (3 per student)
Thin-stem pipettes (for hydrogen peroxide)
Microtip pipettes (for potassium permanganate)
pH test paper strips (for cleanup)

Reagents

Amounts given are more than enough for a class of 24 students working individually. See "Notes on the Procedure."

Commercial hydrogen peroxide, H_2O_2, nominal 3%
6 *M* sulfuric acid, H_2SO_4, in thin-stem pipettes. *Slowly and with constant stirring,* add 30 mL of concentrated acid to 60 mL of distilled water. This concentration is not critical. Wear safety goggles and gloves. The reaction is very exothermic; if possible, clamp the mixing vessel securely in a container of ice water. If cooling is not possible, add the concentrated acid in small batches, separated by periods of cooling.

Potassium permanganate solution, $KMnO_4$. Place 2.1 to 2.3 g (±0.001 g) in a tared beaker. Add 100 mL of distilled water, and record the mass of the solution. Stir for at least 30 minutes to dissolve, and transfer to one or more capped containers as dispensers. Include on the labels the concentration in moles of $KMnO_4$ per gram of solution.

Saturated sodium carbonate solution, Na_2CO_3 (for cleanup) (about 25 g/100 mL)

Discussion

This is not a sophisticated gravimetric titration. The titrant, potassium permanganate, is used without standardization (unless you choose to do otherwise). The experiment is quick and easy and gives very good, reproducible results if students maintain the acidity of the system and agitate the system continually during titration.

Potassium permanganate dissolves slowly. Make this solution first, and then allow it to stir (magnetic stirrer) while you prepare the sulfuric acid and set out other apparatus.

Discuss with students the nature of the products. Oxygen gas can be seen coming from the mixture. Constant swirling is needed to keep it from initiating side reactions, which can affect the pH of the system.

An interesting extension is to prepare some unknowns. Starting with 6% peroxide, make solutions that vary from 2% to 6%. The student directions include two other suggestions.

If you wish to use this system, but with the added step of standardizing the potassium permanganate, students can follow the procedure given for Experiment 20, "Analysis By Oxidation-Reduction Titration." You can adjust the strength of the oxalate given there to match the concentration of the permanganate solution you prepare for this peroxide analysis.

Notes on the Procedure

1. Dispense the peroxide by placing small commercial bottles at convenient locations around the laboratory. Bottles keep well enough if refrigerated between classes (unless classes are consecutive, in which case the bottles need only be capped).

2. It is best to prefill the sulfuric acid pipettes, to prevent acid burns and drippings on the tables. To avoid confusion between the two clear, colorless reagents in the thin-stem pipettes, we find it helpful to trim the stems of the sulfuric acid pipettes by about 3 cm. This also improves students' ability to control where the acid goes.

3. Sample calculations of permanganate concentration: Assuming 2.153 g of $KMnO_4$ is used to make 101.225 g of solution,

   2.153 g $KMnO_4$ ÷ 158.0 g/mol = 0.0136(2658) mol $KMnO_4$
   0.0136(2658) mol $KMnO_4$/101.225 g solution = 1.346×10^{-4} mol $KMnO_4$/g solution

   If the permanganate solution is to be used within a few hours of preparation, it need not be standardized for this experiment.

4. Small Erlenmeyer flasks or beakers are much to be preferred over test tubes for this experiment. The concentration of the permanganate is such that each titration will take nearly one full pipette of titrant.

Answers to Prelaboratory Questions

1. Addition of excess $KMnO_4$ would lead one to report a higher percentage of hydrogen peroxide. In addition, students should recognize that the permanganate excess is magnified 2.5 times, according to the stoichiometry of the reaction.

2. Students often predict that the percentage would fall to zero or nearly so. In fact, due in part to the presence of stabilizers (which also may consume permanganate in the titration), the percentage is usually very close to 3. One of us (JGL) has had students run the experiment with peroxide over 2 years beyond the date and found that there was still just under 2% of the peroxide present.

Conclusion

1. Student answers, but values typically run in the neighborhood of 3.2% to 3.5%.

Answers to Postlaboratory Questions

1. To the extent that such reactions occur, the mass of permanganate solution consumed will increase, leading to a falsely high value for the peroxide concentration. This coincides with the second Prelaboratory Question.

2. Molecular: $H_2SO_4 + Na_2CO_3(aq) \rightarrow H_2O(l) + CO_2(g) + Na_2SO_4(aq)$;
net ionic: $2H^+ + CO_3^{2-}(aq) \rightarrow H_2O(l) + CO_2(g)$.

EXPERIMENT

22 The Oxidation States of Nitrogen

Placement: Oxidation-reduction or descriptive discussion of nitrogen chemistry

Approximate time for completion: 45 minutes

Approximate preparation time: 30 minutes

Materials: 24 of each, unless noted otherwise, assuming a class of 24 students working individually

Apparatus

Milligram balance
Microtip Beral pipettes, 72 (3 per student)
Electrically heated sand bath or other heat source (at least 6; can be shared)
10-mL Erlenmeyer flasks, 72 (3 per student)

Reagents

Amounts given are more than enough for a class of 24 students working individually. See "Notes on the Procedure."

Iron(III) solution, $Fe^{3+}(aq)$. Dissolve 12 g of ferric ammonium sulfate dodecahydrate, $FeNH_4(SO_4)_2{\cdot}12H_2O$ (molar mass 482.2 g), in a small beaker with 20 mL of 6 *M* H_2SO_4; dilute the solution to 100 mL with distilled water. An excess of iron is needed; this solution will give you a 7-fold excess when mixed with an equal volume of hydroxylamine solution. A similar strength solution can be made using 6.7 g of iron(III) chloride hexahydrate, $FeCl_3{\cdot}6H_2O$ (molar mass 270.30 g). Replace the sulfuric acid with 20 mL of 6 *M* HCl.

Hydroxylamine solution, $NH_2OH{\cdot}HCl$. Weigh an empty 150-mL beaker. Add between 0.60 and 0.70 g of hydroxylamine hydrochloride (molar mass 69.5 g). Add about 100 mL distilled water. Weigh the beaker and solution. To determine the concentration of hydroxylamine hydrochloride in moles per gram of solution, calculate moles of hydroxylamine hydrochloride used and divide by the mass of the solution. The amounts given here will produce a concentration of about 8.6×10^{-5} mol/g; students will need the actual concentration.

Potassium permanganate solution, $KMnO_4(aq)$. Prepare in a similar manner to that used for the hydroxylamine hydrochloride solution, but use between 0.30 and 0.40 g $KMnO_4$ (molar mass 158.04 g) and add about 100 mL of water; determine the total mass of the mixture. This gives a solution that has a concentration of about 2.5×10^{-5} mol/g. (The potassium permanganate solution need not be standardized if used within a few days of preparation.)

3 *M* sulfuric acid, H_2SO_4. If stock solution is not available, *slowly and with constant stirring,* add 20 mL of concentrated acid to 100 mL of distilled water. Allow the solution to cool before use.

Discussion

In this experiment, interpretation and manipulation of the data will present far more of a challenge than performing the actual experiment. Students use an excess of iron(III) chloride to oxidize hydroxylamine (in the form of the hydrochloride, sometimes called *hydroxylammonium chloride*). Concentration of the hydroxylamine solution is known.

The iron(II) ions produced are reoxidized to iron(III) via quantitative titration with potassium permanganate. the number of moles of iron(II) oxidized in the titration by permanganate must equal the number of moles of iron(III) that reacted with the hydroxylamine initially.

The number of moles of hydroxylamine that reacted can be found from knowledge of the mass of solution used and the solution concentration, given in moles of NH_2OH per gram of hydroxylamine solution.

Since the iron only changed oxidation state by one (3+ to 2+), the ratio of the moles of iron(III) reduced to the moles of hydroxylamine oxidized must equal the change in the oxidation state for the nitrogen. This makes it a simple matter to determine the final oxidation state for the nitrogen, and the correct product is selected from the options given.

Students should understand that the value of the ratio is to be rounded to the nearest whole number, since the ratio represents the number of electrons lost by the nitrogen atom in hydroxylamine, and fractional electrons do not exist.

Notes on the Procedure

1. The concentration of iron(III) solution is not critical and need not be determined. Iron(III) chloride also can be used, but ferric ammonium sulfate solutions have a longer shelf life.

2. Purchase hydroxylamine as the hydrochloride, $NH_2OH{\cdot}HCl$ (also known as *hydroxylammonium chloride*). Hydroxylamine hydrochloride has a limited shelf life, so it is best to purchase fresh each year and only in the quantities that you expect to consume. Happily, one of us (JGL) has found that even if the solution is prepared with

old reagent, the results obtained by students, while not "correct," were still quite consistent, and good agreement on the identity of the product was obtained. The expected product is nitrous oxide, N_2O.

3. Students will need the concentrations of permanganate and hydroxylamine (hydroxylammonium chloride) solutions. If a milligram balance is used to prepare them, each solution's concentration can be given to three significant digits, although the permanganate concentration will not really merit that degree of precision without being standardized. Standardization does not materially improve the results of the experiment.

4. Hot plates can be substituted for the sand baths; we assume that the sand baths are always available and hence specify them here. If hot plates are used, they can be shared easily by students.

Answers to Prelaboratory Questions

1. Sulfuric acid is used only to maintain a low pH; it has no stoichiometric significance in the system.

2. 1–

3. (1) Iron(III) is the oxidizing agent; hydroxylamine is the reducing agent.
   (2) Permanganate is the oxidizing agent; iron(II) is the reducing agent.

Calculations

5. $8H^+ + 5Fe^{2+} + MnO_4^- \rightarrow 5Fe^{3+} + Mn^{2+} + 4H_2O$

6. N_2O

7. $2NH_3OH^+ + 4Fe^{3+} \rightarrow Fe^{2+} + N_2O + H_2O + 6H^+$

8. The ratio represents the relative numbers of electrons gained and lost by the two ions. A fractional ratio would suggest fractions of electrons, which do not exist.

9. None; water is already the solvent. The only case where it would matter is one in which there was significant iron present in the water.

10. a. It has to be a reducing agent to change iron(III) to iron(II).
    b. No; all the possibilities named have either positive or zero oxidation states for the nitrogen atom(s).
    c. Yes; in ammonia, for example, nitrogen is in the 3– state. This represents a reduction starting from the 1– state of NH_2OH. The other reagent(s) would have to be capable of being oxidized; this rules out Fe^{3+}, for example.

EXPERIMENT

23 Microelectrolysis of a Salt

Placement: Electrochemistry (also can be placed with redox)

Approximate time for completion: 75 minutes (slightly more if students make their own U-tubes

Approximate preparation time: 45 minutes

Materials: 24 of each, unless noted otherwise, assuming a class of 24 students working individually

Apparatus

Glass U-tubes, 4–8 mm o.d.
9-V batteries
Battery clips, bare wire, for Part 1
Battery clips, with alligator clips, for Parts 2–5
Pencil lead electrodes (see "Notes on the Procedure")
Steel wool (0000) or very fine sandpaper or garnet paper for cleaning electrode tips between experiments
Distilled water in wash bottles for rinsing U-tubes between experiments

Reagents

Amounts given are more than enough for a class of 24 students working individually (see "Notes on the Procedure"). The list of reagents below includes all the substances used in the experiment. Not all are listed in the table of reagents in the text.

The solutions are quick and easy to make, and since each student uses only 1 to 2 mL, directions are for making 50 mL of solution. Add the mass of solute given below to 50 mL of distilled water and stir or swirl to mix. Thin-stem Beral-type pipettes work very well for filling the tubes.

0.5 *M* potassium iodide, $KI(aq)$, 4.15 g $KI(s)$
0.5 *M* copper(II) chloride, $CuCl_2(aq)$, 4.26 g $CuCl_2{\cdot}2H_2O$
0.5 *M* nickel(II) nitrate, $Ni(NO_3)_2(aq)$, 7.27 g $Ni(NO_3)_2{\cdot}6H_2O$
0.5 *M* zinc bromide, $ZnBr_2(aq)$, 5.63 g $ZnBr_2$ (see "Notes on the Procedure")
0.5 *M* sodium chloride, $NaCl(aq)$, 1.46 g NaCl
0.5 *M* table salt (iodized sodium chloride), 1.46 g iodized table salt
Phenolphthalein indicator, 0.4% solution
Starch solution. Spray from a pressurized can of spray starch into about 100 mL of distilled water until the mixture has a distinct opalescent blue color. Allow foam to subside before use.

Discussion

This highly simplified procedure is nonetheless a very effective teaching tool. In fact, the electrolysis of KI contains some very sophisticated chemistry that is more than a challenge for beginning students (including the fact that potassium is more active than hydrogen and so is not plated in Part 1).

Nickel, copper, and zinc will be plated out. Part 5 is intended to show students that table salt is not pure sodium chloride. That the concentrations of solutions are not particularly critical is reflected in the directions for solution preparation under "Reagents" above.

Notes on the Procedure

1. The text directs students to run the electrolysis of KI for 5 minutes. Generally, this will be far longer than is necessary to get results; 2 minutes or so may be enough in this and the succeeding parts, although it sometimes takes a while to get enough of the metals to see in Parts 2 to 4.

2. As the procedure suggests, alligator clips on the battery leads are not necessary in Part 1. There is no reason, however, that students cannot do the entire experiment with draftsmen's leads as electrodes.

   Any mechanical drawing leads with preformed points will do, although harder leads, such as the 2-mm HB drawing leads, are generally longer lasting. The polymer-based leads tend to soften with use, so they cannot be used by more than one or two sections.

   In Parts 2 to 4, metal will be plated on the tip of the cathode each time. Students will need to either scrape the tip with a sharp knife or (better) polish the tip clean with fine garnet paper or sandpaper or 0000 steel wool.

3. Be sure that the table salt is iodized. Just to make the point, you may wish to make a box of (iodized) table salt available and have students prepare their own solution for Step 9.

4. The choice of zinc bromide is relatively arbitrary. The crystals of zinc are the goal here, so unless you want to have a third halogen involved, you can substitute zinc chloride, 3.41 g.

5. While it isn't included in the text, you might want to provide students with short (10-cm) lengths of 4- to 6-mm glass tubing and have them make their own U-tubes. If so, they will need gas burners and wing tops. The glass can be placed on ceramic plates or wire screens or leaned against the base of a ring stand to cool. Since this may well be their first experience with bending glass, give a demonstration of proper technique during the prelaboratory discussion and expect about 25% waste.

Answers to Prelaboratory Questions

1. Sodium hydroxide solution and chlorine gas.

2. They are needed to provide plating surfaces for the metals in Parts 2 to 4. In Part 5 they simply provide a good surface for oxidation and reduction.

Data and Observations

Part 1

1. Dark blue color appears (starch/iodine complex).

2. Iodine must have been produced at the negative electrode (the anode).

3. Negative. $2I^- \rightarrow I_2 + 2e$. Electrons are lost (oxidation).

4. a. No
   b. Water
   c. i. Negative
      ii. Hydroxide
      iii. Hydrogen ions, H^+
      iv. Hydrogen gas, H_2
      v. $2H^+ + 2e \rightarrow H_2$
   d. Combustion

Parts 2, 3, and 4

5. Copper and nickel are seen as thin veneers on the point of the electrode. Zinc often will form small crystal nodules. All equations are of the form $M^{2+} + 2e \rightarrow M$. In each case, the metal is observed at the negative electrode (cathode). All are reductions.

6. In addition to the odors, students may note a straw yellow in the case of bromine. All equations are of the form $2X^- \rightarrow X_2$, and all occur at the positive electrode (cathode). All are oxidations.

7. a. $Cu^{2+}(aq) + 2Cl^-(aq) \rightarrow Cu(s) + Cl_2(g)$. This is the net ionic equation for the electrolysis and includes both the oxidation and reduction half-reactions.
   b. $Zn^{2+}(aq) + 2Br^-(aq) \rightarrow Zn(s) + Br_2(g)$

8. Students rarely see chlorine's faint color, but bromine and iodine are pretty easy to spot.

9. A few students may notice the color difference at the anode (positive electrode) due to the formation of small amounts of molecular iodine. Most will not.

10. Yes, there is a somewhat faint but definitely positive test for iodine.

11. The table salt will be labeled as "iodized."

EXPERIMENT

24 Calorimetry: Heat of Fusion of Ice, Heat of Neutralization, Specific Heat of a Metal, and Heat of Solution

Placement: Thermodynamics; if text has early treatment of calorimetry and enthalpy with later chapter covering entropy and free energy, place this with the earlier chapter.

Approximate time for completion: 2 to 2.5 hours if all parts are done

Approximate preparation time: 30 minutes

Materials: 24 of each, unless noted otherwise, assuming a class of 24 students working individually

Apparatus

Milligram or centigram balance (1 per 8 students)
Thermometers (mercury- or alcohol-filled, digital, or computer-interfaced probes)
Pipettes, 5-mL or similar, graduated (Mohr)
Graduated cylinders, 25-mL (1 per student for Part 1, 2 per student for Part 4 desirable)
Graduated cylinders, 50- or 100-mL (1 per student for Parts 2, 3, and 5)
Styrofoam cups, 6 oz, 72 (3 per student)
Beakers, 250-mL (supports for calorimeters)
Hot water bath

Reagents

Amounts given are more than enough for a class of 24 students working individually (see "Notes on the Procedure").

Distilled water
Ice, crushed or fine pieces (if necessary, crush large pieces by wrapping in towel and hitting with mallet)
Metal shot, lead, copper, nickel, tin, etc. (Pieces should not exceed 1 mm diameter)

2.00 *M* $HCl(aq)$. *Slowly and with constant stirring,* add 163 mL of concentrated (12.3 *M*) acid to 500 mL distilled water. Dilute to 1.00 L. *Caution: Large amounts of heat are released; wear safety goggles and allow solution to cool before use.*

2.00 *M* $NaOH(aq)$. Dissolve 80.0 g NaOH pellets in water. Stir to dissolve. *Caution: Large amounts of heat are released; wear safety goggles and allow solution to cool before use.* Dilute to 1.00 L at room temperature.

Anhydrous salts for Part 5: calcium chloride, sodium or potassium nitrate, ammonium chloride or nitrate, etc. About 0.050 mol of solute per student.

Discussion

If the problems in your text consistently tell students to neglect the heat capacity of the calorimeter, you can, at your option, shorten the experiment significantly by skipping Part 1, but we feel that calculation of the "calorimeter constant" is valuable both as an experimental activity and in terms of using the result in later calculations. The choice of glass versus digital thermometers versus interfaced probes is more a matter of what you have at hand than value of results. None of the three gives significantly better results, although students tend to have more faith in digital thermometers and are less likely to misread them. The computer-interfaced probes are a convenient and elegant way to collect data, and students can have data plotted by the software supplied with the interfacing unit. The student procedure gives instructions for using the Vernier software; if you have different hardware and software, you will want to cover it in the prelaboratory discussion.

Notes on the Procedure

1. One way to get consistent results in making the calorimeter lid is to put a rubber band around the cup about one-third of the way down from the mouth. Students can either mark the line, remove the rubber band, and make the cut or they can simply use the rubber band as a guide. Disposable wallpaper knives are inexpensive cutting tools.

2. For many students, understanding the nature of the calculations in Part 1 presents a problem. For this reason, we have provided specific instructions. Even so, coverage in both the prelaboratory and postlaboratory discussions may be helpful to students.

3. Students may be amused by the directions in Part 2 to (a) allow the ice to warm up and (b) dry it just before it is added to the water in the calorimeter. Explain the reason for each part; surprisingly enough, they do make a difference in some cases.

4. Small shot (1–3 mm dia.) is preferable to metal cylinders in that the shot seems to heat up more effectively. In either case, the efficiency with which the transfer is made is the key to good results.

5. During the prelaboratory discussion, remind students of the need to carefully dry the inside of the calorimeter between trials and to try to maintain the snug fit for the thermometer or probe.

Answers to Prelaboratory Questions

1. Water is known to have a density of 1.0 g/mL, allowing conversion from milliliters to grams.

2. 1 J/g·°C × (453.6 g/1 lb) × (5°C/9°F) = 252 J/lb·°F = 1 Btu

3. a. q = 100 g × 4.184 J/g·°C × (26.9 – 20.0) = 2886.96 = 2900 J (2 significant figures)
   b. 2900 J; lacking knowledge of the specific heat of the metal, q must be assumed to be the same for both water and metal.
   c. 2886.96 J ÷ [22.50 g × (100.0°C – 26.9°C)] = 1.76 J/g·°C (3 significant figures are justified here.)

Answer to Postlaboratory Question

1. Student data; the procedure introduced with Experiment 1 is used.

EXPERIMENT

25 Thermodynamics of the Oxidation of Acetone by Hypochlorite

Placement: Calorimetry or thermodynamics (see "Notes on the Procedure")

Approximate time for completion: 1 hour

Approximate preparation time: 20 minutes

Materials: 12 of each, unless noted otherwise, assuming a class of 24 students working in pairs

Apparatus

Beakers, 50-mL
Thermometer, –10 to 110°C or thermistor probe for computer-interfacing (see notes for Experiment 24 for discussion of using computer-interfaced probe)
5-mL graduated (Mohr) pipette
25-mL graduated cylinder
6-oz Styrofoam cups, 3 per team (8-oz is OK)
Beakers, 250-mL (optional—to serve as bases for the calorimeters, per Experiment 24)

Reagents

Amounts given are more than enough for a class of 24 students working individually. See "Notes on the Procedure."

Distilled water
1 L household bleach, 5.25% NaOCl (Clorox, Purex, or store brand, unscented)
5.0% aqueous acetone. Dissolve 25.0 mL of acetone in distilled water to make 500 mL of solution.

Discussion

As pointed out in connection with Experiment 24, each of these experiments can be used as a stand-alone, or they can be done at different points in the course if the text being used has an early experiment on enthalpy, followed later by a more extensive discussion of thermodynamics. In the latter situation, we recommend placing Experiment 24 early on, with the present experiment being done much later, due to the more sophisticated reaction system involved.

The change in calorimeter apparatus is due to the use of acetone. One of us (JGL) also has had good results with an alternate construction in which a 50-mL beaker is wrapped with a paper towel and then nested snugly inside a 100- or 150-mL beaker. The lid is an untrimmed 6-oz expanded polystyrene cup drilled to receive the thermometer or probe.

It is necessary to point out that chloroform is produced in the reaction. Students must be made aware of the special handling that is required; please emphasize the "Cleaning Up" instructions during the prelaboratory instructions.

Notes on the Procedure

1. As is the case with most kinetics and calorimetry experiments, one pair of hands is going to find this a difficult experiment to manipulate; for this reason, we recommend having students work in pairs. This is especially true if you are using computer-interfaced temperature probes, rather than ordinary thermometers. This experiment lends itself nicely to use of the temperature probes, such as the Vernier software and hardware. One of us (KLW) has experimented extensively with the Vernier products with great success.

2. Setting the specific heat of solutions equal to that of pure water with 4 significant figures (4.184 J/g·°C) may be overstating the case, but we consider it a good tradeoff in having students deal with significant figures. This might be an excellent topic for postlaboratory discussions. Point out that if the 5.00 mL has an uncertainty of ± 0.05 mL, this is a 1% uncertainty; 1% of 4.184 = 0.042, so the best we can demand of the specific heat is 4.18 ± 0.04.

3. Even though the data table for the specific heat determination precedes that for the determination of ΔH for the oxidation of acetone by bleach, you will notice that we have students carry out the ΔH determination first. There are two reasons for this. First, we felt that it was the better way to avoid having residual heat from one part affect the results of the other. Second, we wanted to give the instructor the option to dispense with the specific heat determination, especially since C_p is negligibly small with respect to the value of ΔH for this particular system.

4. If standard-bore thermometers are used, a no. 2 cork borer makes a good, snugly fitting hole in the bottom of the expanded polystyrene cup. No matter how carefully the lids are handled, after the several trials called for in the procedure, the hole will become widened enough that the cups cannot be passed along from one laboratory section to the next.

5. See Experiment 24 for discussions of replacing the thermometer with thermistor probes and interfacing the experiment with computers.

Answers to Prelaboratory Questions

1. $Cl_2(aq) + 2NaOH(aq) \rightarrow 2NaOCl(aq) + H_2(g)$ or
   $Cl_2(aq) + 2OH^-(aq) \rightarrow 2OCl^-(aq) + H_2(g)$

2. This may be copied directly from the Introduction:
   $OCl^-(aq) + (CH_3)_2CO(aq) \rightarrow CH_3COO^-(aq) + CHCl_3(l)$

3. Yes; one Celsius degree and one Kelvin represent the same temperature interval.

Calculations

1. a. $q = 24.0 \text{ g} \times 4.184 \times \Delta T$ for all three.

2. a. $(0.05 \times 4.0 \text{ mL} \times 0.791 \text{ g/mL}) \div (60. \text{ g/mol}) = 2.64 \times 10^{-3}$ mol
   b. Expect values about 400 kJ/mol.

5. Answers will vary, but typical values for the heat capacity are 20 to 22 J/K, and ΔT values for the oxidations run about 10 K, so the contribution of the heat capacity is about 400 to 450 J. Since the values for *a* (observed) are about 1000 to 1100 J, on average, the contribution of the calorimeter is quite significant.

6. By definition, $\Delta E = \Delta H + P\Delta V$. Since the system is open, there is no way to obtain a value for ΔV. However, if the ideal gas law is employed, we can replace the work term $P\Delta V$ by $\Delta n(\text{gas})RT$, and one could argue that no work is done [$\Delta n(\text{gas}) = 0$], so $\Delta E = \Delta H$.

EXPERIMENT

26 Reaction Kinetics: The Aldol Condensation of Acetone with Benzaldehyde

Placement: Kinetics

Approximate time for completion: 2 hours

Approximate preparation time: 30 minutes

Materials: 24 of each, unless noted otherwise, assuming a class of 24 students working individually

Apparatus

Test tubes, 13 × 100 mm
Pipettes, 1.0 mL
Pipette pumps, or rubber bulb to fill pipette
Beakers, 250 or 400 mL, for water bath
Ice
Hot water or hot plate or electrically heated sand bath (at least 12)
Thermometers
Glass stirring rods
Beakers, 100 mL, for waste
Funnels and filter paper to collect crystals
Stopwatches

Reagents

1 *M* acetone, 8.7 g (11.0 mL), made up to 150 mL with ethanol
2 *M* benzaldehyde, 31.8 g (30.6 mL), made up to 150 mL with ethanol; 36 g dissolved in water and made up to 300 mL with water
3*M* sodium hydroxide. Dissolve 120 g of sodium hydroxide in 500 mL of water in a large beaker with stirring. *Caution: The solution will become very warm.* After it cools to room temperature, dilute it to 1 L.

95% ethanol
Ice
3*M* hydrochloric acid. Slowly and with stirring, add 75 mL of concentrated (12.3 *M*) reagent to 225 mL of distilled water. Allow solution to cool before use.
Hot water

Discussion

The bottles of acetone and benzaldehyde can be labeled directly with their concentrations (1 *M* acetone and 2 *M* benzaldehyde in 95% ethanol) or with the weights per milliliter or the volumes per milliliter, in which case the students will need to calculate the concentrations themselves. The densities of acetone and benzaldehyde are 0.790 and 1.04 g/mL, respectively.

These quantities of reagents will given enough material to carry out 150 reactions. If 15 students carry out the nine reactions in Parts 1 and 2A and 2B, then 135 mL of the acetone and benzaldehyde solutions will be needed. If the effects of concentration are studied, then more material will be needed.

Ordinary technical-quality acetone can be used. If the benzaldehyde is old and shows the presence of benzoic acid, put it in a beaker and stir it with sodium bicarbonate solution until gas evolution stops. Then pour the mixture into a separatory funnel and remove the lower benzaldehyde layer. Store it over calcium chloride to remove any droplets of water. Do not try to distill the benzaldehyde; it oxidizes very readily during this process unless done under an inert atmosphere.

It is advantageous to have pairs of students carry out this experiment. One can mix the solutions, and the other can do the timing. Depending on the time available, the latter reactions do not need to be run in triplicate.

For the purposes of a general chemistry course, this reaction can be regarded simply as $2A + B \rightarrow C + 2D$, where we are measuring the rate of appearance of C. It is, of course, not quite this simple. First of all, A reacts with B to form an intermediate I, which then reacts with another molecule of A to form C. From the standpoint of organic chemistry, two moles of benzaldehyde react with one of acetone to form the bright yellow crystalline dibenzalacetone.

The mechanism of the reaction involves first the formation of the enolate anion under the influence of the hydroxide ion catalyst. This enolate anion then attacks the carbonyl of the benzaldehyde to form an unstable alcohol that spontaneously loses the elements of water to form benzalacetone as an intermediate. Once all the acetone is consumed, the process is repeated with the enolate anion of the intermediate attacking a second molecule of benzaldehyde to eventually form the product. It is the appearance of this bright yellow product as it comes out of solution that we can use as the endpoint of the reaction.

It is somewhat remarkable that the product forms in such high yield, because a number of side reactions can be visualized; for example, the enolate anion of the acetone attacks another molecule of acetone, or it attacks the carbonyl of the benzalacetone.

If you play a bit with molecular models, you will readily see that the double bonds in the product can, theoretically, be either *cis* or *trans*. In fact the *trans-trans* product is the only one formed. That yellow color is due to the long path of alternating double and single bonds that leads to low-energy electronic transitions.

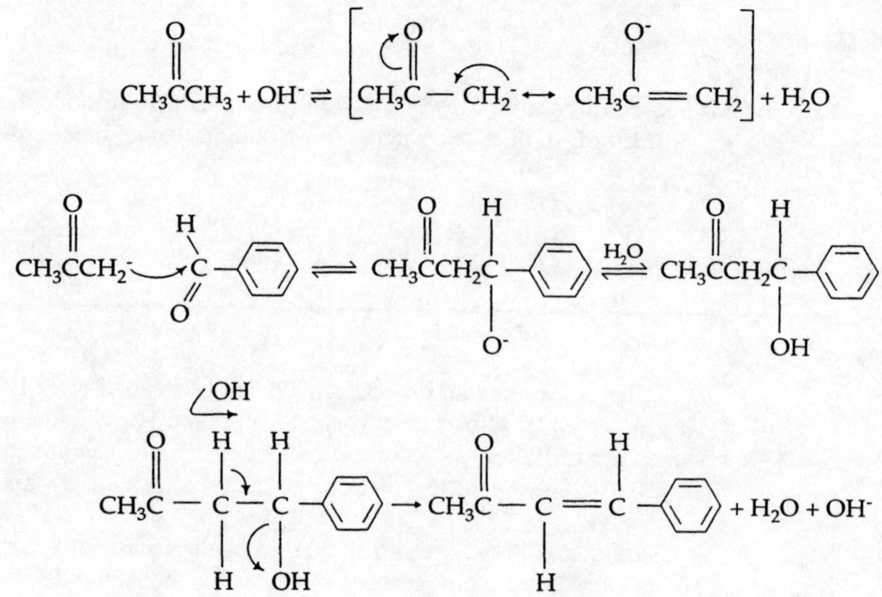

Safety: Although the Aldrich catalog lists benzaldehyde as a toxic irritant, it usually causes no problems in the open laboratory, although good ventilation is mandatory. It is sold in the grocery store in alcoholic solution as "artificial almond flavor." Three molar sodium hydroxide solution is hazardous to the skin and particularly the eyes.

Safety goggles must be worn at all times. If some of the solution gets on the skin, it should be washed with copious amounts of water until the skin no longer feels slippery. Provide soap and water so that students can wash their hands during and after the experiments. If the solution gets in the eyes, flushing with running water for 15 minutes is necessary. Call a physician. There should be no open flames present because acetone, benzaldehyde, and ethanol are all flammable.

Cleaning Up: All the reaction mixtures from a given pair of students can be placed in a beaker. This material can be filtered through fluted paper to remove the dibenzalacetone and washed with a little ethanol. The filtrate can then be neutralized with 3 *M* hydrochloric acid and flushed down the drain with water. The dry dibenzalacetone is not toxic and can be disposed of in the trash. It gives very large, nice crystals when recrystallized from ethanol.

Notes on the Procedure

If time is limited, you may have students do either Part 2A or Part 2B. More time and reagents will be needed to study the effect of concentration as described in Parts 2C and 2D. Because this experiment is applicable at many levels, no one set of questions would be appropriate to all students.

Answers to Prelaboratory Questions

1. Answers will vary. Oxidation reactions are the most obvious. These would include burning, decomposition of animal and vegetable matter (hence refrigeration), and so on.

2. A catalyst is, by definition, a substance that is unchanged in a chemical reaction. The concentration of hydroxide ion does not change as a function of time. It is a constant and thus does not appear in the rate equation.

EXPERIMENT

27 Metal Reactivities

Placement: Descriptive chemistry, periodic table, or in connection with the chapter on oxidation and reduction

Approximate time for completion: 20 minutes (pair with Experiment 28, "Nonmetal Reactivities")

Approximate preparation time: 45 minutes

Materials: 24 of each, unless noted otherwise, assuming a class of 24 students working individually

Apparatus

Safety goggles
Laboratory apron and gloves (optional)
24-well test plate
Forceps or tweezers, for handling small metal samples
Wash bottle and cotton swabs (for cleanup)
Hand lens (optional, but useful for detecting faint changes)

Reagents

Unless noted otherwise, amounts are for a class of 24 students working individually. Directions for preparation are for 100 mL of solution, enough for at least 6 sections.

Test solutions in microtip pipettes. All solutions are prepared by dissolving the indicated mass of solute in 100 mL of distilled water.
0.2 *M* $CuCl_2(aq)$, 3.4 g of the dihydrate
0.2 *M* $FeCl_3(aq)$, 5.4 g of the hexahydrate
0.2 *M* $Mg(NO_3)_2(aq)$, 5.1 g of the hexahydrate

0.2 M $SnCl_4(aq)$, 7.0 g of the pentahydrate
0.2 M $Zn(NO_3)_2(aq)$, 6.0 g of the hexahydrate

While it is the usual practice to add a few drops of acid to solutions of transition metals to stabilize them against reaction with the air, to do so would be counterproductive here, since the protons of the acids (HCl, HNO_3) will oxidize three of the metals being tested.

Small pieces (about 2 × 2 mm) of

Cu, Sn, Zn. The pieces of Cu, Sn, and Zn are conveniently prepared by cutting up strips or sheets of 0.010-in-thick foil.

Fe. Changes are more easily observed with the steel wool. For iron, use pieces of clean wire or small balls of 0000 steel wool that have been rinsed free of their oil coating.

Mg. Magnesium pieces can be cut directly from clean ribbon.

Discussion

The experiment is done quickly. We have revisited the concept of net ionic equations versus molecular equations in both the discussion and the Postlaboratory Questions as a refresher. As part of the Prelaboratory discussion, reinforce the connection between activities of metals and their relative ease of oxidation but do not let students push it too far; the $E°$ values in the table of standard reduction potentials are not in 100% accord with observed activities.

A good follow-up is to demonstrate the reaction between aqueous silver nitrate and metallic copper. (This demonstration is referred to in one of the Postlaboratory Questions but is not included in the procedure. If you elect not to do the demonstration, direct students to omit Postlaboratory Question 4.

The demonstration: A loose coil of oil-free copper wire is prepared by wrapping the wire (18 to 22 gauge) around a pencil, leaving part of it straight. The straight end is inserted through the hole in a no. 4 rubber stopper, and the assembly is lowered into a 25 × 150 mm test tube containing 0.1 M $AgNO_3(aq)$. Crystals of silver begin to form immediately, becoming longer and more lustrous as the reaction proceeds. The tube can be clamped in position and left overnight or longer. Along with formation of crystalline silver, the blue color characteristic of copper(II) ions becomes evident in the solution. A more elegant version of the demonstration can be done with a projecting microscope or microvideo camera by placing a 1- to 2-cm length of very thin copper wire in a few drops of silver nitrate solution on a slide or Petri dish. Formation of the crystals is seen more easily, but the color change of the solution is not as easily detected.

Notes on the Procedure

Little can go wrong here unless students ignore the instruction to observe changes over a reasonable length of time. Some of the changes are subtle, hence the suggestion that the well plate be left in a safe place for later examination. If you combine this experiment with Experiment 28, "Nonmetal Reactivities," as suggested above, students can make their preliminary observations of this experiment and then cover the well plate (to prevent contamination) and set it aside while they do Experiment 28. Final observations can then be made after cleanup of the second experiment.

Answer to Prelaboratory Question

1. Gold is the least reactive metal and virtually impervious to its environment. It will not form a nonconducting oxide coating on its surface. It also, of course, is an excellent electrical conductor.

Analysis and Conclusions

1. Observations: Mg reacts with all other cation solutions; it is the most active of the five.

2. Copper reacts with none, so it is the least active.

3. Mg > Zn > Fe > Sn > Cu. Detection of the tin reaction with cupric chloride is not easy. Reaction is faint and slow, resulting in some students ranking tin and copper equally. If steel wool is used for iron, and if it is not rinsed free of its protective oil coating, detection of iron's reactivity is difficult.

4. Since the metallic copper is clearly oxidized by silver ions, copper must be the more active of the two, placing silver last on the list for the experiment.

Answers to Postlaboratory Questions

1. Aluminum is more active because it can be oxidized by cupric ions.

2. No; these reactions are not spontaneously reversible. Students have not yet encountered the concepts of free energy and electromotive force, but they have the experience on which to base their decision.

3. Reactions that did occur and whose numbers should be circled: b, d, f, g, h. Reactions that did not occur and so should be crossed out: a, c, e.

4. Nickel is less active than magnesium and iron (and therefore zinc, based on the experiment) but more reactive than copper and tin (and silver, from the demonstration). This places nickel between iron and tin: Mg > Zn > Fe > Ni > Sn > Cu > Ag.

5. The anion with which zinc ions pair up must be whatever anion was present in the solution to which zinc was added. (In fact, Zn^{2+} will be a hydrated ion, not part of a compound, in both examples.)

6. $Ni(s) + CuCl_2(aq) \rightarrow NiCl_2(aq) + Cu(s)$ or $Ni(s) + Cu^{2+}(aq) \rightarrow Ni^{2+}(aq) + Cu(s)$; $2Ni(s) + SnCl_4(aq) \rightarrow 2NiCl_2(aq) + Sn(s)$ or $2Ni(s) + Sn^{4+}(aq) \rightarrow 2Ni^{2+}(aq) + Sn$

7. a. To inhibit the oxidizing effect of the air.
   b. The oil also would resist oxidation by the cations of less active metals, such as copper.
   c. Water would not work; oils are immiscible with water.
   d. The more active the metal, the more such a coating is necessary, so it would be more effective for metals near the start of the activity series.

EXPERIMENT

28 Nonmetal Reactivities

Placement: Descriptive chemistry, the periodic table, oxidation-reduction; makes a natural pair with Experiment 27, "Metal Reactivities"

Approximate time for completion: 40 minutes

Approximate preparation time: 30 minutes

Materials: 24 of each, unless noted otherwise, assuming a class of 24 students working individually

Apparatus

Reaction tubes or 10 × 100-mm culture tubes (ideally, 6 to 8 per student, but a smaller number can be cleaned and reused periodically)
Corks to fit tubes (these, too, can be reused without significant contamination)

Reagents

Directions are given for preparing 100 mL of each solution.

Chlorine water. In a fume hood, slowly add 5 mL of 3 *M* H_2SO_4 to 20 mL of household bleach, 5.25% of NaOCl(*aq*). Dilute to 100 mL. Store in dark glass bottle.
Bromine water. Dissolve 0.042 g $KBrO_3$ in 25 mL distilled water. Add 0.72 g KBr and swirl to dissolve. Slowly add 6 mL of 3 *M* H_2SO_4 with gentle swirling. Dilute the yellow-brown solution to 100 mL. Store in a dark glass bottle.
Iodine water. Dissolve 0.054 g KIO_3 in 25 mL distilled water. Add 1.0 g KI and swirl to dissolve. Cautiously add 6 mL of 3 *M* H_2SO_4 with gentle swirling. Dilute the dark-red solution to 100 mL.
150 mL *tert*-butyl methyl ether, *t*-bme
1.0 *M* sodium bromide, 10.3 g NaBr

1.0 *M* sodium chloride, 5.9 g NaCl
1.0 *M* sodium iodide, 15.0 g NaI
Sodium sulfite, solid, 24 g
Sodium thiosulfate solution, 1 *M* $Na_2S_2O_3$, in dropping bottles (for cleanup), 15.8 g anhydrous solid or 24.8 g of the pentahydrate in 100 mL of distilled water

Discussion

The halogens are the only nonmetals that can be used conveniently to extend the activity concept. The molecular halogens are toxic; adequate ventilation and protective clothing are definitely indicated, as is thoroughness of cleanup.

If you want to minimize the amounts of molecular halogens present in the laboratory, the experiment works quite well with 0.5 *M* halide ion solutions; results start to get less consistent when solution concentrations are reduced to 0.2 *M*.

The bromine-, chlorine-, and iodine-water solutions produced by the directions above are significantly more concentrated than commercially available solutions and give very distinctive colors.

Special disposal considerations are involved for the *t*-bme, but the use of sodium thiosulfate to convert molecular halogens to halide ions is both instructive and cost-efficient. If students have studied or are studying the chapter on oxidation-reduction reactions, writing balanced equations for the halogen to halide conversions makes a good exercise. We did not include them in the student manual in order to allow more flexibility of placement.

This experiment is the natural companion to Experiment 27, "Metal Reactivities."

Notes on the Procedure

1. The "Preliminary Tests" are intended to show students what color the molecular halogens give to the organic (*t*-bme) layer. It will help immensely if you reinforce this point during prelaboratory discussions and as students work.

2. Another point to make clear during prelaboratory discussions is that the *-ine* ending (chlorine, bromine, iodine) refers specifically to the molecular halogens, while *-ide* (chloride, etc.) can only be the anion form. Even by midyear, students have a tendency to use *-ine* to refer to either form, thus heightening their own confusion over which species is where and at what point in the procedure.

3. Corks are not watertight, so some moisture will leak through. A small piece of plastic wrap or "dental dam" rubber sheeting will help to keep the solutions off students' skin and clothing.

4. Because chlorine is the most active of the three molecular halogens tested, students will see evidence that it converts bromide to bromine (yellow-brown *t*-bme layer) and iodide to iodine (reddish *t*-bme layer). Bromine reacts with iodide but not chloride, and iodine water reacts with none of the halide solutions.

Analysis of Results

There are two questions at the end of this section, neither of which has a blank on the report sheet.

1. Molecular fluorine would be expected to react with all the other halide ion solutions.

2. Fluorine is so reactive that it will vigorously attack anything—including water—that it can oxidize.

Answers to Prelaboratory Questions

1. Chlorine: pale yellow-green gas. Bromine: dark red liquid. Iodine: intensely dark purple solid.

2. Fluorine is gaining electrons, so something in the water must be losing them. The hydrogen is still there and has no electron to lose anyway, so the only possibility is oxygen. In fact, fluorine is the only element that is sufficiently active to convert oxide ions to oxygen gas: $2F_2(g) + 2H_2O(l) \rightarrow 4HF(aq) + O_2(g)$.

Data and Observations

Part 1

chlorine: colorless
bromine: yellow-orange
iodine: reddish

Part 2

Activity tests with chlorine water and *t*-bme
chloride: n.r.
bromide: yellow-orange organic layer; reaction: Br_2 forms
iodide: red organic layer; reaction: I_2 forms
Activities with bromine water and *t*-bme
chloride: n.r.
bromide: n.r.
iodide: blue-black
Activities with iodine water and *t*-bme
chloride: n.r.
bromide: n.r.
iodide: red organic layer; not a reaction: I_2 was added
Color produced on addition of starch solution to
chlorine: n.r.
bromine: n.r.
iodine: blue-black

Appearance of solutions after addition of starch solution to
chloride ion: n.r.
bromide ion: n.r.
iodide ion: some color in both layers due to solubility of I_2 and I_3^- in both solvents

Conclusions

1. Of the three tested, chlorine is the most reactive and iodine is least reactive. By inference, the pattern appears to show that fluorine is most active of the four halogens. This fits the predictions of periodicity, since one would expect fluoride to demonstrate the highest degree of electron affinity due to its small radius.

Answers to Postlaboratory Questions

1. Note the wording, ". . . all reactions that *occurred.*"
   $Cl_2(aq) + 2Br^-(aq) \rightarrow 2Cl^-(aq) + Br_2(aq)$
   $Cl_2(aq) + 2I^-(aq) \rightarrow 2Cl^-(aq) + I_2(aq)$
   $Br_2(aq) + 2I^-(aq) \rightarrow 2Br^-(aq) + I_2(aq)$

2. Only molecular iodine reacts with the starch solution.

EXPERIMENT

29 Synthesis and Quantitative Analysis of an Iron Compound

Placement: Late in the course; involves redox titration, spectrophotometric analysis, and volumetric dilutions; suitable as a final laboratory project

Approximate time for completion: 2 to 3 laboratory periods

Approximate preparation time: 1 hour

Materials: 24 of each, unless noted otherwise, assuming a class of 24 students working individually

Apparatus

Milligram or analytical balances (at least 6)
Reaction tubes or 100-mm test tubes (not culture tubes) (48, 2 per student)
Electrically heated sand bath or hot plate or hot water bath
Boiling sticks
Microtip pipettes (96, 4 per student)
25-mL volumetric flask
Spectronic 20-D or similar spectrophotometer (4 or more per class)
Tubes or cuvettes for use with spectrophotometer
Drying oven, adjusted to 110 to 120°C (enough oven space for 24 samples)
Desiccator (adequate space for 24 samples)
10-mL Erlenmeyer flasks or similar small titration vessels (72, 3 per student)

Reagents

Amounts of solids allow for spillage, repeats. Solution preparation directions provide more than enough for one class of 24 individuals.

Iron(III) chloride hexahydrate, $FeCl_3{\cdot}6H_2O$, 10 g

Potassium oxalate monohydrate, $K_2C_2O_4{\cdot}H_2O$, 20 g
Acetone (wash bottles)
Distilled water (wash bottles)
Ice, crushed (for inducing crystallization in reaction mixtures)
0.2 *M* sulfuric acid, H_2SO_4, 600 mL (25 mL per student). Slowly and with constant stirring, add 11 mL of concentrated acid to 989 mL of distilled water. *Caution: The process is highly exothermic. Too-rapid addition of the acid can cause eruption. Wear protective clothing and work in a fume hood.*
6 *M* sulfuric acid, H_2SO_4. Slowly and with constant stirring, add 33 mL of concentrated acid to 67 mL of distilled water. *Caution: The process is highly exothermic. Too-rapid addition of the acid can cause eruption. Pause occasionally to allow the system to cool (or support mixing vessel in a large container of ice water). Wear protective clothing and work in a fume hood.*
Potassium permanganate solution, $KMnO_4$ (about 2×10^{-5} mol $KMnO_4$/g solution. In a tared container, dissolve 0.27 to 0.3 g (±1 mg) of solid in 100 mL of distilled water. Reweigh container and solution. This gives a concentration of about 2.2×10^{-5} mol $KMnO_4$/g of solution. If the solution is prepared freshly and with care, the concentration is sufficiently precise that students need not standardize it further. See "Notes on the Procedure."

Note: Students must be given the concentration of the solution to 3 significant figures.

Standard iron(II) solution, 0.0250 mg Fe^{2+}/mL. Dissolve 1.755 g ferrous ammonium sulfate hexahydrate, $Fe(NH_4)_2(SO_4)_2{\cdot}6H_2O$, in a small amount of water in a 1-L volumetric flask, with 2.5 mL of concentrated sulfuric acid, H_2SO_4 (*caution*). When the solid has dissolved, the flask is filled to the mark with distilled water and the contents mixed thoroughly. This solution has a concentration of 0.250 mg Fe^{2+}/mL. If you wish to complete the solution preparation for your students, continue with the next paragraph.

Using a volumetric pipette, transfer exactly 10.00 mL to a clean 100-mL volumetric flask, and dilute to the mark with distilled water. Again, mix the contents thoroughly by inverting the flask a number of times. This solution has the desired 0.0250 mg Fe^{2+}/mL concentration.

Ammonium acetate, $NH_4C_2H_3O_2$, 100 g (see "Notes on the Procedure")
Hydroxylamine hydrochloride, $H_2NOH{\cdot}HCl$, 100 g (see "Notes on the Procedure")
1,10-phenanthroline, 10 g (see "Notes on the Procedure")

Discussion

This experiment is a scaled-down version of a similar activity found in many laboratory manuals. It has been reduced by a factor of 30 below the typical quantities of material yet still gives decent results while introducing students to use of volumetric glassware and a spectrophotometer. The chemistry involved is fairly elegant. Iron(III) is precipitated as Iron(III) potassium oxalate trihydrate, $K_3Fe(C_2O_4)_3{\cdot}3H_2O$. The product is well known, appearing in the *CRC Handbook of Chemistry and Physics*, as well as other references. It loses its waters of hydration at 100°C and has a solubility of less than 5 g/100 cc H_2O at 0°C. While the procedure given here does not call for determination of waters of hydration (due to the small amount of product obtained), it would be reasonable to have some members of a class determine the percentage of water while others did the analyses for iron and oxalate.

Alternatively, all students could carry out the complete analysis using the dried product for preparation of the solution that is used for both the iron and oxalate determinations.

If you wish to have your students carry out the water determination, be aware that the compound decomposes at 230°C, so dehydration will require a drying oven. A small (10- to 20-mL) beaker or flask might be a good container for this purpose.

Typical data appear below:

	Theoretical Percentage	Average Student Results
Iron	11.4%	10.5%
Oxalate	53.8%	54%
Water	11.0%	11%

Notes on the Procedure

1. The permanganate solution is the same one that is used in Experiment 20, "Analysis by Oxidation-Reduction Titration." If you desire to have students standardize the solution, you can refer them to that experiment for instructions.

2. Even though it is included in the "Procedure" section of the experiment, you should prepare the initial ferrous ammonium sulfate solution, as described above. Each student in the section can then withdraw the 10.00 mL of this solution and prepare his or her own 0.025 mg/mL solution.

3. You can reduce costs—and time—significantly by preparing the solutions of ammonium acetate, hydroxylammonium chloride, 1,10-phenanthroline, and the final preparation of the iron(II) standard in advance. This also reduces the amounts of glassware needed. If you choose to make the first three, the quantities given in the procedure (3.85 g ammonium acetate, $NH_4C_2H_3O_2$, per 50 mL, 3.48 g hydroxylamine hydrochloride, NH_3OHCl, per 50 mL, and 0.3 g 1,10-phenanthroline, $C_{12}H_8N_2$, per 100 mL) will be enough for an entire section of 24 students.

Answers to Prelaboratory Questions

1. 0.250 mg/mL = 0.250 g/L; 0.250 g Fe/L × (392.16 g $Fe(NH_4)_2SO_4{\cdot}6H_2O$/55.85 g Fe) = 1.755 g $Fe(NH_4)_2SO_4{\cdot}6H_2O$ needed, so you would dissolve that mass of solid reagent in sufficient water to make 1.00 L of solution. Assuming the solution has a density essentially equal to that of water, 1.00 g/mL, there are 0.250×10^{-3} g of Fe per gram of solution or 250 g Fe per million grams of solution, hence 250 ppm Fe.

2. $16H^+ + 2MnO_4^- + 5C_2O_4^{2-} \rightarrow 10CO_2 + 2Mn^{2+} + 8H_2O$

3. Thermal decomposition, including loss of waters of hydration.

Answers to Postlaboratory Questions

1. Student answers.

2. $H_2C_2O_4^{2+} + 2Fe^{3+} \rightarrow 2CO_2 + 2Fe^{2+}$; $E°$ for the oxalate half-cell is available in some references but not the *CRC Handbook of Chemistry and Physics*. The most likely sources are quantitative analysis texts. $E°$ for oxalate/carbon dioxide is 0.49 V, and $E°$ for iron(III)/iron(II) is 0.77 V, so $E°$ for the cell is 1.26 V. The reaction is thermodynamically favored.

3. a. Mol oxalate per mol iron(III) = 3.52:1, using the masses suggested in the procedure.
   b. Easily. The reduction requires only 0.5 mol oxalate per mole of iron, and the formation of the iron(II) oxalate requires an additional mole of oxalate; hence the stoichiometric minimum is 1.5:1.

4. The molar mass of the compound is 491.26 g; this contains 117.30 g K, 55.85 g Fe, 264.06 g $C_2O_4^{2-}$, and 54.06 g H_2O. The percentages are K = 23.88%, Fe = 11.37%, $C_2O_4^{2-}$ = 53.75%, and H_2O = 11.00%.

EXPERIMENT

30 Partition Coefficient of an Organic Acid

Placement: Acid-base titrations, solubility and solutions

Approximate time for completion: 1 to 1.5 hours

Approximate preparation time: 20 minutes

Materials: 24 of each, unless noted otherwise, assuming a class of 24 students working individually

Apparatus

Reaction tubes or 10 × 100-mm culture tubes or test tubes
Corks or stoppers to fit tubes
Microtip pipettes or Pasteur pipettes (for removing aqueous layer following extraction)
10-mL Erlenmeyer flasks or other titration vessels, 3 (or more) per student, if possible
Mohr (graduated) pipettes (2 per student), for measuring volumes of acetic acid and *t*-bme for final threefold extraction

Reagents

All but the *t*-bme are dispensed in labeled microtip pipettes. Provide several small stock bottles in convenient locations for students to refill pipettes as needed. Phenolphthalein pipettes will not need refilling.

Acetic acid, 1.00 *M*. Slowly and with constant stirring, add 11.5 mL of glacial reagent to 188.5 mL of distilled water.
Sodium hydroxide, 1.00 *M*. Dissolve 8 g of solid sodium hydroxide pellets in water to make 200 mL of solution. Standardize by gravimetric titration; give students the concentration of the base in moles of NaOH per gram of solution.
Phenophthalein indicator solution, 0.1%

Discussion

The experiment uses familiar skills to investigate further the properties of solutions and the effect of molecular polarity on solubility. Titrations are gravimetric.

Notes on the Procedure

1. Reasonable results have been obtained when NaOH was treated as a primary standard. Thus 8.00 g of good-condition NaOH pellets is weighed (±0.01 g) into a tared beaker; 200 mL of distilled water is added, and the solution is weighed again. Even though the concentrations are not correct, the value for the partition coefficient is acceptable.

2. Propanoic acid is a better choice if hexane, TTE, or some other entirely nonpolar solvent is used, but its solubility in *t*-bme is too high for use here.

3. Students may already have experience at inserting a plastic or Pasteur pipette into a tube containing liquid without drawing unwanted material into the tip, but remind them to exert a constant gentle pressure on the bulb during insertion. Also emphasize the need to avoid drawing in liquid from the interface or from the organic layer.

4. The marking on the reaction tubes are sufficiently precise for the combination of 1.00 mL each of the acid and ether called for in the final step of the procedure. Rather than using Mohr pipettes, students can be directed to prepare the mixture by mass. The tube is placed in a tared beaker on the balance. Acid is added from the pipette until the mass reaches exactly 1.00 g; then *t*-bme is added from a separate pipette until the mass has risen by 1.00 g.

Answer to Prelaboratory Question

1. a. (1.01 g HOAc/mL) × (1.00 mL *t*-bme/0.740 g *t*-bme) = 1.36 mL *t-bme*
   b. 1.00 g *t*-bme × (1.00 mL/0.740 g *t*-bme) = 1.35 mL *t*-bme

Conclusions

1. a. Extraction 1: $1.5 = x/(10 - x)$; $15 - 1.5x = x$; $2.5x = 15$; $x = 6$ mmol of acid in the top layer, which is 1 mL, or 0.74 g, using the density of *t*-bme as 0.74 g/mL. (6 mmol/g) × (0.74 g) gives us 4.4 mmol of acid in the top layer. This means that the bottom (aqueous) layer retains 25.6 mmol of acid, for a concentration of 8.5 mmol/mL.

   Extraction 2: $1.5 = x'/(8.5 - x')$; $12.75 - 1.5x' = x'$; $2.5x' = 12.75$; $x' = 5.1$ mmol of acid. The bottom layer retains (25.6 – 5.1 =) 20.4 mmol, for a concentration of 6.8 mmol/mL.

Extraction 3: $1.5 = x''/(6.8 - x'')$; $10.2 - 1.5x'' = x''$; $2.5x'' = 10.2x'' = 4.08$ mmol of acid. The bottom layer retains (20.4 – 4.1 =) 16.3 mmol of acid, for a final concentration of 5.4 mmol/mL or 5.4 mmol/g.

b. All is the same as before, except the volume of organic layer is increased threefold. It still has a concentration of 6 mmol/g, but it now contains (6 mmol/g) × (0.74 g/mL) × 3 mL = 13.3 mmol of acid. This leaves (30 – 13.3 =) 16.7 mmol of acid in the bottom layer, for a concentration of 5.8 mmol/g. The aqueous layer begins with 3 mL, 10 mmol/g, or 30 mmol of acid.

2. $1.5 = x/10 - x$; if the concentration of acid in the organic layer is 98.5% of the initial concentration of 10 mmol/g, then the value of $10 - x$ must be 0.15. Similarly, the number of millimoles of acid removed is (9.85 mmol/g) × 3 g = 29.55 mmol. $1.5 = x/0.15$; $x = 0.225$ mmol/g *t*-bme; (0.225 mmol acid/g *t*-bme) × (mass of *t*-bme needed) = 29.55 mmol of acid; mass of *t*-bme needed = 29.55 ÷ 0.225 = 131 g (177 mL) of *t*-bme.

EXPERIMENT

31 Analysis of Commercial Vinegar

Placement: Acids and bases

Approximate time for completion: 1.5 hours

Approximate preparation time: 30 minutes

Materials: 24 of each, unless noted otherwise, assuming a class of 24 students working individually

Apparatus

Milligram balances (at least 1 for every 8 students)
50-mL beakers or flasks, for preparing KHP solution
Microtip pipettes (3 per student)
10-mL Erlenmeyer flasks or small (20–30 mL) beakers (3 or more per student)
Volumetric or Mohr pipette, 2.5-mL capacity, with bulb or syringe for filling 25-mL volumetric flasks

Reagents

Amounts given are more than enough for a class of 24 students working individually. See "Notes on the Procedure."

Potassium hydrogen phthalate, 15 g, oven dried at 105 to 110°C. See "Notes on the Procedure."
Distilled water
Sodium hydroxide solution, NaOH(*aq*), approximately 0.1 *M*. Dissolve 0.40 g of solid in 100 mL of water that has been boiled briefly to remove dissolved carbon dioxide.
Phenolphthalein solution, 0.5%, in 50:50 alcohol and water
"White" vinegar, about 75 mL. See "Notes on the Procedure."

Discussion

Except for the use of gravimetric rather than volumetric analysis, this is a very familiar experiment. We have chosen to present it in a relatively simple fashion, but you can easily expand on what is here or even change the focus somewhat by making more of the fact that the equivalence point of a strong base–weak acid titration has a pH above 7. You can simplify the experiment quite a lot by providing a prepared KHP solution (see note 2 of "Notes on the Procedure") or by eliminating the dilution of the vinegar. The latter option will mean that a stronger (~1 *M*) NaOH solution will be needed.

If you have them available, pH meters, even the hand-held units that sell for about $40, work very nicely in this system. If used with the indicator, they even show that the pH is not 7 at equivalence. This concept is particularly appropriate for making clear the concepts of hydrolysis and strong versus weak as they apply to acids and bases.

This experiment also can be done quite nicely with computer-interfaced probes. The materials available from Vernier give good results and are very student friendly.

Notes on the Procedure

1. A "quick and dirty" variation can be done with the hand-held portable pH meters mentioned above, using only a 24-well test plate and nonstandardized NaOH. Place 15 drops of acid, either ~1.0 *M* HCl or vinegar (~0.83 *M* HOAc), in a well. Add 1.0 *M* NaOH dropwise, gently stirring with the pH meter and noting the pH after each drop. Even on the small scale used here, the difference in the shape of the pH curve for strong and weak acids will be evident.

2. You may elect to prepare the potassium acid phthalate standard yourself; this will save about 30 minutes of student laboratory time and reduce waste of materials. Dry the solid at 105 to 110°C, and then weigh out 2.04 g for each 100 mL of solution to be prepared. The 100 mL will handle about 24 students working individually, allowing for small amounts of waste.

3. In Part II you might prefer to have students dilute 5 mL of vinegar to 50 mL, especially if you have 5-mL volumetric pipettes. While more vinegar is consumed and wasted, cost is minimal, and there is no hazard or disposal problem.

4. Three or four small (1-pt) bottles of grocery store (house brand" vinegar placed at convenient locations around the room work very well. While this is not the most efficient way to provide "stock" solutions, it helps to remind students of the connection between "chemicals" and real-life materials.

Answers to Prelaboratory Questions

1. The only hydrogen that is different from the others on the structure is the one connected to the carboxyl group; it must be the acidic hydrogen, since if one of the others were acidic, then all four would have to be.

2. (0.0250 L) × (0.100 mol KHP/L) × (204.23 g KHP/mol) = 0.510575 = 0.511 g KHP.

3. Since volumes, densities, and molarities are equal for the two, so must the masses be: 1.00 g of NaOH solution will be needed.

4. Calculations are the same for both: (0.100 mol/L) × (1 L/1000 mL) × (1.00 mL/1.00 g) = 1.00×10^{-4} mol/g of solution.

5. a. (5 g HOAc/100 g solution) × (1000 g/1.0L) × (1 mol HOAc/60.0 g HOAc) = 0.833 mol HOAc/L = 0.8 *M*.
   b. Mol HOAc = mol NaOH; 1.0 mL HOAc × (0.8 mmol HOAc/mL) = V_{NaOH} × (0.1 mmol NaOH/L); V_{NaOH} = 8 mL.
   c. No; 8 mL would be about 8 g of solution. The pipette would have to be refilled two times, possibly three.

Calculations and Conclusions

1. For each trial: {(mass of KHP solution used) × (conc. of KHP in mol/g)} ÷ (mass of NaOH solution used) = concentration of NaOH. Students calculate mean value, individual deviation for each trial, and average deviation in the usual manner.

2. a. Concentration of diluted vinegar is found in the manner described for NaOH above.
   b. The original vinegar was 10 times as strong as the diluted one, so the calculation of mass percent works like this:

      {[(mol HOAc/g dilute vinegar) × 10.00 × (60.0 g HOAc/mol)] ÷ mass of dilute vinegar used} × 100% = mass percentage of acetic acid

      Again, the values are averaged, a mean is found, and average deviation is determined.

Analysis and Results

1. a. (i) $CO_2(aq) + OH^-(aq) \Leftrightarrow HCO_3^-(aq)$
      (ii) $HCO_3^-(aq) + OH^-(aq) \Leftrightarrow CO_3^{2-}(aq) + H_2O(l)$
   b. Without side reactions, the concentration will be 0.100 *M* in NaOH. Hopefully, students will recognize that the bicarbonate (and carbonate) ions produced are still bases, albeit less strong. In principle, then, the neutralizing power is the same, since acetic acid is a stronger acid than carbonic acid.

2. Since the carbon dioxide reacts with the hydroxide ions, the solubility must be increased; this is an application of Le Châtelier's principle.

3. a. Boiling the water drives off CO_2, preventing the side reactions described above.
   b. As noted in the answer to Question 1, the ultimate effect should be negligible, although carbonate and bicarbonate are weaker bases than hydroxide.

4. $Ca^{2+}(aq) + CO_3^{2-}(aq) \rightarrow CaCO_3(s)$; $Mg^{2+}(aq) + CO_3^{2-}(aq) \rightarrow MgCO_3(s)$.

5. As noted in the Prelaboratory Questions, large amounts of the same NaOH solution would be needed; refilling the pipette would introduce greater experimental uncertainty in the result. This could be dealt with by using approximately 1 *M* NaOH.

6. a. $HC_8H_4O_4^-$ (base) and $H_2C_8H_4O_4$ (acid), OH^- (base) and H_2O (acid)
   b. $HC_2H_3O_2$ (acid) and $C_2H_3O_2^-$ (base); OH^- (base) and H_2O (acid)

EXPERIMENT

32 Gas Chemistry: Carbon Dioxide

Placement: Descriptive chemistry, gas chemistry

Approximate time for completion: 2 hours

Approximate preparation time: 30 minutes

Materials: 24 of each, unless noted otherwise, assuming a class of 24 students working individually

Apparatus

The gas generator is made by drilling slightly undersize holes in the caps of 3-dram vials and then forcing heavy-walled polyethylene tubes through the holes. The polyethylene tubing is wrapped on a wood mandrel and dipped in boiling water to permanently shape the bends in the tubing before cutting it to length (see figure).

Reagents

Bromthymol blue indicator
Calcium carbonate
3 *M* hydrochloric acid
Saturated calcium hydroxide solution
Saturated barium hydroxide solution

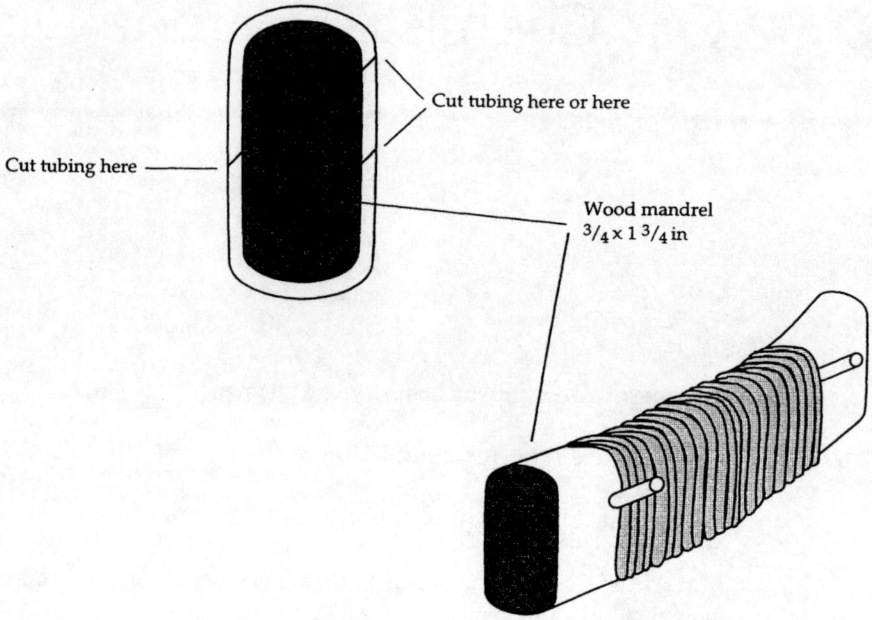

Mandrel for shaping polyethylene tubing. The pieces of tubing can have either both arms of the same length or one short and one long arm.

Discussion

In vial D the precipitate of calcium carbonate that is formed may redissolve if enough carbon dioxide passes through the solution. The insoluble calcium carbonate is converted to soluble calcium bicarbonate.

Cleaning Up: The procedure says to empty the vials into the waste container in the hood and then rinse out all vials, tubes, and pipettes with water; however, the products from this experiment are innocuous and, except for the solids, could be washed down the drain.

Answer to Prelaboratory Question

1. The apparatus would not blow up if the tube leading from vial B to vial C became blocked. The pressure would be relieved through the safety tube in vial A, which has a hole cut into it at the top. Under ordinary circumstances, the bottom of the safety tube is under the liquid in vial A so that gas goes on through the apparatus.

Calculations and Conclusions

1. Three molar hydrochloric acid contains three moles per thousand milliters (a liter), so one milliter contains 0.003 mol of HCl.

2. From the equation for the reaction it is seen that 2 moles of hydrochloric acid will react with 1 mole of calcium carbonate, so 1.5 millimoles of calcium carbonate will react with 3 millimoles of hydrochloric acid.

3. From question 2 it has been established that 1.5 mmol of calcium carbonate is needed. The molecular weight of calcium carbonate is 100.09 g, so

   $0.0015 \text{ mol} \times 100.09 \text{ g/mol} = 0.15 \text{ g}$

   Thus 150 mg of calcium carbonate is required.

4. From the equation and question 3 it is seen that the number of moles of carbon dioxide is equal to the number of moles of calcium carbonate that will react with the hydrochloric acid. This is 1.5 mmol. In this reaction the hydrochloric acid is the *limiting reagent*. The calcium chloride, 200 mg, is in excess.

5. At standard temperature and pressure (STP), one mole of a gas occupies 22,400 mL, so 0.0015 mol would occupy 33.6 mL.

6. Under very high pressure, carbon dioxide can be observed to be a liquid at room temperature.

7. The mole percent of argon in air is much higher than the mole percent of carbon dioxide in air.

8. The substance formed is calcium carbonate and is made from the reaction of the hydroxide with carbon dioxide in the air:

   $Ca(OH)_2 + CO_2 \rightarrow CaCO_3 + H_2O$

EXPERIMENT

33 Gas Chemistry: Ammonia

Placement: Descriptive chemistry, gas chemistry

Approximate time for completion: 2 hours

Approximate preparation time: 30 minutes

Materials: 24 of each, unless noted otherwise, assuming a class of 24 students working individually

Apparatus

Gas generation and reaction apparatus

Reagents

Ammonium chloride
Calcium hydroxide
Universal indicator
1 *M* $CoCl_2$ solution, 23.80 g/100 mL H_2O of the hexahydrate
1 *M* $NiCl_2$ solution, 12.96 g/100 mL H_2O
1 *M* $CuSO_4$ solution, 24.97 g/100 mL H_2O of the pentahydrate
Concentrated hydrochloric acid

Answers to Prelaboratory Questions

1. The apparatus would not blow up if the tube leading from vial B to vial C became blocked. The pressure would be relieved through the safety tube in vial A, which has a hole cut into it at the top. Under ordinary circumstances, the bottom of the safety tube is under the liquid in vial A so that gas goes on through the apparatus.

2. None of the tubes in vials B through F dip below the level of the liquid in the vial, so ammonia has free passage through the apparatus. Enough ammonia is generated so that a change occurs in vial E (the solution turns very dark blue as the copper ammonium complex is generated). The vial of HCl is there to ensure that ammonia does not escape into the room.

Answers to Postlaboratory Questions

1. $NH_3 + H_2O \rightarrow NH^{4+} + OH^-$. Ammonia acting as a Lewis base picks up a proton from water, forming some hydroxide ion, which is basic.

2. See "Introduction."

3. $NH_3(g) + HCl(aq) \rightarrow NH_4Cl(s)$.

EXPERIMENT

34 Gas Chemistry: Sulfur Dioxide

Placement: Physical properties, separation of mixtures

Approximate time for completion: 2 hours

Approximate preparation time: 30 minutes

Materials: 24 of each, unless noted otherwise, assuming a class of 24 students working individually

Apparatus

Gas generation apparatus. See Experiment 32 in this *Instructor's Guide*.

Reagents

Sodium sulfite
Bromcresol green indicator
0.01 *M* potassium permanganate solution. Dissolve 0.74 g of potassium permanganate in 500 mL of water.
3 *M* sulfuric acid
2 *M* sodium hydroxide solution. Dissolve 40 g of sodium hydroxide in water and make the solution up to 500 mL.
Potassium dichromate
3 *M* hydrochloric acid

Discussion

In vial C the purple permanganate will turn colorless or very light pink as $Mn^{2+}(aq)$ is formed. In vial D manganese dioxide is formed. This brown solid is formed in such small amounts that it may be very difficult to see. In vial E green $MnO_4^{2-}(aq)$ is formed from the purple permanganate. In vial F green $Cr^{3+}(aq)$ is formed from the orange dichromate. In vial G the appearance of the charcoal is unchanged.

Notes on the Procedure

If the special projects are performed, one can expect the following changes in color of vanadanate:

$VO_3^-(aq) \rightarrow VO^{2+}$ yellow to green

$VO^{2+} \rightarrow V^{3+}$ green to blue

$V^{3+} \rightarrow V^{2+}$ blue to violet

Iodine is reduced to iodide by sulfur dioxide:

$I_2 + SO_2 + H_2O \rightarrow 2\ HI + SO_3$

Answers to Prelaboratory Questions

1. The balanced equation for the reaction of acidified dichromate ion with sulfur dioxide in aqueous solution is

   $Cr_2O_7^{2-}(aq) + 3SO_2(g) + 5H^+(aq) \rightarrow 2Cr^{3+}(aq) + 3HSO_4^-(aq) + H_2O\ (l)$

2. $SO_2(g) + O_3(g) \rightarrow SO_3(g) + O_2(g)$
   $SO_3(g) + H_2O(l) \rightarrow H_2SO_4(aq)$

Physical Properties of Reagents

Molecular weight of sodium sulfite: 126.04 g/mol
Solubility of sodium sulfite: 12.5 g/100 mL of water at 0°C
Molecular weight of sulfur dioxide: 64.06 g/mol
Density of sulfur dioxide gas: 2.927 g/L
Color of potassium permanganate, $KMnO_4$: dark purple
Color of potassium manganate, K_2MnO_4: green
Color of manganese chloride, iodide, or bromide, MnX_2: pink
Color of manganese dioxide, MnO_2 (insoluble in water): brown

Answers to Postlaboratory Questions

1. The amount of sodium sulfite is 0.40 g/126.04 g/mol = 0.0032 mol = 3.2 mmol.

2. The acid contains 6 mol/L or 6 mmol/mL.

3. From the first equation it is seen that 2 mol of hydrochloric acid reacts with 1 mole of sodium sulfite to give 1 mol of sulfur dioxide, so 6 mmol of acid will react with 3 mmol of sulfite (even though 3.2 mmol is present) to give 3 mmol of sulfur dioxide. Hydrochloric acid is the limiting reagent in this reaction. One mole of a gas at STP occupies 22,400 mL, so 0.003 mol occupies 22,400 mL/mol × 0.003 mol = 67.2 mL.

EXPERIMENT

35 Determination of Molar Mass by Vapor Density and Freezing Point Depression

Placement: Properties of solutions, colligative properties (also reviews gas laws)

Approximate time for completion: 1 to 1.5 hours, depending on number of trials run

Approximate preparation time: 20 minutes (plus instructional time spent on use of computer interface equipment, if needed)

Materials: 24 of each, unless noted otherwise, assuming a class of 24 students working individually

Apparatus

Part 1

Milligram balance (at least 1 for every 6 students)
100-mL round-bottom flask
Flask heater (100-mL Thermowell) and controller
Aluminum foil, pieces 5 cm^2, plus larger squares (~20 cm^2) as heating "tents"
Copper wire
250-mL graduated cylinders (see "Notes on the Procedure")

Part 2

Thermometers, digital, or digital thermometer probe for computer interface
Reaction tubes (or 10 × 100-mm test tubes or culture tubes)
Sand for flask heaters (freezing-point depression method only)
Safety goggles

Reagents

Amounts given are more than enough for a class of 24 students working individually. See "Notes on the Procedure."

Part 1

Unknown volatile liquids, about 5 mL per student. Here are some suggestions:

Liquid	Boiling Point (°C)	
Acetone	56.2	
Methanol	65	
Hexane(s)	69	(mixed isomers work well enough)
Ethanol	78.5	
2-Butanone	79.6	
2-Propanol	82.4	
1-Propanol	97.4	
2-Butanol	99.5	

Part 2

Biphenyl, about 25 g
Unknown solid, 100 mg per student. Use any low-polarity molecular solids.
 Anthracene
 Fluorene
 p-Dichlorobenzene
 Naphthalene

Discussion

The experiment is placed near discussions of colligative properties, but many of the calculations recall chapters on the gas laws. Both methods have historical importance, although the freezing-point depression technique gives more precise results. It is important that the solute chosen for Part 2 be something that will not associate to any great extent; notwithstanding the potential for hydrogen bonding, urea gives good results.

Notes on the Procedure

1. Nominal 100-mL round-bottom flasks (nonstandard taper) have a volume of 110 to 120 mL. A 100-mL graduate may be used for determining the volume of the round-bottom flask, but it will need to be filled to the 100-mL mark, emptied, and filled again. This introduces extra uncertainty into the calculations, hence the specification of a larger size.

2. We have had excellent results with the Vernier software available directly from Vernier or from other suppliers such as Flinn Scientific and PASCO Scientific. The Vernier

materials can be used with both Macintosh and IBM computers and are also designed for use with Texas Instruments 80 series graphing calculators. Other systems are available, and all are sufficiently similar that there will be little difference in the student directions.

3. As shown in Figure 35.2 of the student text, if supercooling occurs, students should be instructed to extrapolate back along the postsupercooled incline to the region of the cooling curve that precedes supercooling. The temperature at which the extrapolated line meets the initial cooling curve is taken as the freezing point.

Answers to Prelaboratory Questions

1. $n = PV/RT$.

2. MW = m/n.

3. MW = $m/(PV/RT) = mRT/PV$.

Answers to Postlaboratory Questions

Part 1: Molar Mass by Vapor Density Method

1. Since MW is directly proportional to n and n is directly proportional to T, a 10% error in temperature results in a 10% error in the molecular weight. The experiment is being carried out at a presumed 400 K; this allows a 40 K (40°C) leeway.

2. The argument here is the same as for question 1. The leeway now is 0.1 atm = 76 torr = 10 kPa.

3. Given the disclaimer in the "Introduction," the assumptions are quite valid for the level of precision we are trying to attain.

4. The system is open to the surroundings; no buoyancy is involved.

5. The vapor of the unknown is very near its boiling temperature, so it will be far from ideal, since the molecules are occupying a significant volume and intermolecular forces are certainly important.

6. The final mass of flask and contents will be too high, since both condensed vapor and unvaporized sample will be present. The result will be an elevated MW value.

Part 2: Molar Mass by Freezing Point Depression

7. The molecular weight is directly proportional to the freezing point depression, so a 10% error in the molecular weight results when the temperature error is 10% of the amount by which the freezing point is lowered. This is not the same as a 10% error in reading the temperature.

EXPERIMENT

36 Solubility, Complex Ions, and Qualitative Analysis

Placement: Chapter on solution equilibria

Approximate time for completion: 2 hours

Approximate preparation time: 45 minutes

Materials: 24 of each, unless noted otherwise, assuming a class of 24 students working individually

Apparatus

96-well test plates
10 × 100-mm test tubes or reaction tubes (144, 6 per student)
10-mL graduated cylinder
100-mL graduated cylinder
250-mL beaker
Safety goggles

Reagents

Cation set in microtip pipettes, 0.20 *M*. Masses shown are for preparing 50 mL of each. This is enough for at least two sections.

Solute	Test Ion	Mass of Reagent
$Na_2C_2O_4$	Na^+	1.34 g
KNO_3	K^+	1.01 g
$Ca(NO_3)_2$	Ca^{2+}	2.36 g of the tetrahydrate
$Fe(NO_3)_3$	Fe^{3+}	4.04 g of the nonahydrate
$Cu(NO_3)_2$	Cu^{2+}	2.42 g of the trihydrate

Solute	Test Ion	Mass of Reagent
$Al(NO_3)_3$	Al^{3+}	3.75 g of the nonahydrate
$Pb(NO_3)_2$	Pb^{2+}	3.31 g
$Zn(NO_3)_2$	Zn^{2+}	2.97 g of the hexahydrate
$Co(NO_3)_2$	Co^{2+}	2.91 g of the hexahydrate
Anion set in microtip pipettes, 0.20 *M*		
$NaC_2H_3O_2$	$C_2H_3O_2^-$	1.36 g of the trihydrate, 0.82 g anhydrous
Na_2SO_4	SO_4^{2-}	1.42 g anhydrous or 3.22 g of decahydrate (Glauber's salt)
NaOH	OH^-	0.40 g
Na_2SO_3	SO_3^{2-}	1.26 g
Na_3PO_4	PO_4^{3-}	3.80 g of the dodecahydrate
Na_2CO_3	CO_3^{2-}	1.06 g anhydrous or 1.24 g monohydrate
NaCl	Cl^-	0.59 g

Other reagents

$AgNO_3$, 0.10 *M*. Dissolve 0.85 g in 50 mL of distilled water.

$NH_3(aq)$, 4 *M*. Slowly stir 27 mL of concentrated reagent into 73 mL distilled water.

$HCl(aq)$, 1 *M*. Slowly stir 8 mL of concentrated reagent into 92 mL of distilled water.

KSCN, 0.10 *M*. Dissolve 0.97 g of solid reagent in 100 mL of distilled water.

$Na_2S_2O_3$, 0.20 *M*. Dissolve 2.48 g of pentahydrate in 100 mL of distilled water.

Starch solution. Spray from a pressurized can of spray starch into about 100 mL of distilled water stirring constantly until the mixture has a distinct opalescent blue color. Allow foam to subside before use.

Bleach, 5.25% NaOCl. Fill pipettes directly from a small, reasonably fresh bottle of commercial bleach (Purex, Clorox, or similar)

Halide solutions, knowns and unknowns. Prepare 100 mL of 0.2 *M* solutions of NaF (0.84 g), NaCl (1.17 g), NaBr (2.06 g), and NaI (3.00 g). Dilute 75 mL of each with 75 mL of distilled water to make the 0.1 *M* "knowns" and the single-halide unknowns (for Step 17). Use the remainder, full-strength, to make the two-halide unknowns for Step 18.

Chlorine water (for Part 4, optional). In a fume hood, slowly add 5 mL of 3 *M* H_2SO_4 to 20 mL of household bleach, 5.25% $NaOCl(aq)$. Dilute to 100 mL. Store in a dark glass bottle.

tert-Butyl methyl ether (*t*-bme), 50 mL

Discussion

This experiment combines several concepts, one of which we felt deserved more treatment than it is usually given: complex ions. Placement in the chapter on equilibria in aqueous solution matches it well with the concepts of K_{sp} for ionic solids and $I_{instability}$ for complexes, while affording the opportunity for quite a lot of refresher practice in writing complete and net ionic equations for reactions between ions in solution. You may, for example, have students write both types of equation for all cases in Part 1 in which a precipitate was observed.

Notes on the Procedure

1. Step 16 is included for completeness, but it does not add measurably to the student's understanding. The chlorine water must be fresh.

2. Demonstrate the technique illustrated by Figure 36.1.

Answers to Prelaboratory Questions

1. The concentrations of the cations are 0.05 *M*; that of chloride is 0.1 *M*.

2. Ammonium chloride, NH_4CL: $NH_3(g) + HCl(g) \rightarrow NH_4CL(s)$.

Answers to Postlaboratory Questions

1. The first drop of 4 *M* NH_3 should neutralize the 4 drops of 1 *M* HCl. If students are careful and mix thoroughly, they almost always find that 5 drops is enough to dissolve the precipitate. They also know, or can figure out, that 10 drops of 0.1 *M* silver nitrate give the same amount of silver ion that is in 1 drop of 1 *M* HCl. Even so, they only know that there are no more than 4 ammonia molecules per silver. While they will not be particularly happy with the uncertainty, it makes an excellent exercise to suggest why a more dilute acid and ammonia solution would not be as acceptable. On the other hand, since the sodium thiosulfate used in Step 8 is 0.2 *M*, it is quite reasonable to expect students to get close to the correct formula for that complex: $Ag(S_2O_3)_2^{3-}$.

2. a. $Sr^{2+}(aq) + 2\,F^-(aq) \rightarrow SrF_2(s)$.
   b. No reaction occurs.
   c. No reaction occurs.
   d. $Ag^+(aq) + Cl^-(aq) \rightarrow AgCl(s)$.

3. Nitric acid converts the ammonia to ammonium ions, freeing them from the silver ions and allowing the precipitate of silver chloride to form once again. This is part of the traditional qualitative analysis scheme for the insoluble chloride group.

4. Especially after standing for a minute or two, the three have distinctive colors. AgCl is white at first, turning gray on standing; AgBr is ivory, white with a yellowish tinge; AgI has a definite yellowish cast, which some students describe as orange.

EXPERIMENT

37 Water Softening

Placement: Solutions, solubility, applied chemistry

Approximate time for completion: 30 to 45 minutes; pairs well with other short laboratories

Approximate preparation time: 40 minutes; reduces to 10 minutes if students fill their own pipettes

Materials: 24 of each, unless noted otherwise, assuming a class of 24 students working individually

Apparatus

Pasteur pipettes (96, 4 per student), each set of 4 containing fine sand (1), ion-exchange resin (1), and cotton plugs (2)

10×100-mm test tubes or reaction tubes, with corks to fit, 96 (8 per student would be better)

Reagents

Amounts given are more than enough for a class of 24 students working individually. See "Notes on the Procedure."

Artificial hard water. A few grams of calcium chloride or sulfate are added to a gallon of water.

Calgon. The Calgon-treated hard water is prepared by adding the appropriate amount of Calgon to a few hundred milliliters of the artificially hardened water. The directions on the box are for 1 gallon of water, about 1500 mL.

Sodium carbonate test solution, 0.1 *M* Na_2CO_3. Add about 0.5 g of anhydrous Na_2CO_3 or 0.6 g of monohydrate to 50 mL of distilled water.

Powdered soap, such as Ivory or castile soap or liquid Ivory. Avoid using detergents such as Tide.

Discussion

This quick and simple experiment makes a nice break near the end of the year using familiar materials and reinforcing concepts.

Notes on the Procedure

1. None of the concentrations is critical. If the water in your region runs over about 150 ppm total hardness, you do not really need to artificially harden the water to be used, but be sure it gives a positive test with sodium carbonate.

2. For filling one of the filter pipettes with ion-exchange resin, make a small funnel by cutting the top of the bulb off a microtip plastic pipette and trimming off all but about 1 to 2 cm of the stem. This will fit snugly on the Pasteur pipette. This is where most of the preparation time goes if you prepare the pipettes. If students are to do their own, you will need to explain how to pour a slurry. These pipettes must be filled a little at a time. A similar funnel can be used for sand.

3. The illustration in the text (Figure 37.1) shows the pipette underfilled. The sand or resin should extend to about 2 cm below the constriction at the top.

4. Emphasize the difference between the turbidity and suds tests, Steps 5 and 6. Point out that they must be done in the prescribed order.

Answer to Prelaboratory Question

1. When hard water evaporates, the calcium salts are left behind. These appear as spots on glassware and shower doors and as the deposit of insoluble salts in the bottom of tea kettles. They also give skin an "unclean" feeling when showering or bathing with true soap and may even cause rust spots on cement areas where there are large iron concentrations in the water.

Conclusions

1. The ion-exchange and Calgon filters do equally well, on average, and the other two do little or nothing.

2. Calgon-treated water would have little value for drinking or cooking. The only real concern for ion exchange is pH if protons are the replacement ion or sodium ions in cases where users are at risk for certain types of heart disease.

3. Sand and cotton alone have nothing in them to attract and hold the calcium and magnesium ions. Their only value is in filtering out relatively large, undissolved contaminants.

4. The more mineral ions there are in the water, the less soap will be able to dissolve and the less effective the cleaning will be. Each mole of calcium ion will cause the precipitation of 2 moles of soap as the insoluble salt. This will decrease the concentration of soap in solution and thus its cleaning ability.

5. Calgon and similar products trap the mineral ions by complexing them. When the Calgon is rinsed away, the mineral ions go with it, so no ring can form.

6. The carbonate removes any mineral ions that might be present, rendering later tests meaningless.

EXPERIMENT

38 Growing Crystals in Gels

Placement: This is a long-running experiment that should be set up at the beginning of a term and observed as the days pass. The gel should be made one week and the other solution added to the top of the gel the next week. These experiments are based on the work of Suib[1] and the Sharbaughs.[2]

Approximate time for completion: 2 hours week 1, 1 hour week 2

Approximate preparation time: 3 to 4 hours

Materials: 24 of each, unless noted otherwise, assuming a class of 24 students working individually

Apparatus

Pasteur pipettes
Pipettes, 1.0 × 0.01 mL
Culture tubes, 10 × 100 mm
Labels for tubes
Corks, to seal tubes

Reagents

Sodium silicate solution. Dilute 154 mL of saturated sodium silicate (density 1.39) to 1 L. Sodium silicate (water glass) is available from most suppliers. The diluted water glass solution should meet the following criteria, although good results can be obtained

[1]S. J. Suib, "Crystal Growth in Gels," *J. Chem. Ed.*, **62**, 81–82 (1985).

[2]A. H. Sharbaugh III and A. H. Sharbaugh Jr., "An Experimental Study of the Liesegang Phenomenon and Crystal Growth in Silica Gels," *J. Chem. Ed.*, **66**, 589–594 (1989).

without these properties: The specific gravity should be 1.06, and the normality should be 0.60 when titrated with standard HCl to a bromothymol blue endpoint.

1.0 *M* acetic acid, CH_3COOH. Mix 57 mL of concentrated (glacial) acetic acid with water to make 1 L. Note that some of the experiments call for 0.5 *M* or 0.6 *M* acetic acid, which the students can prepare.

Metals or ions in solution, as indicated below. This list can be edited to eliminate, for instance, the heavy metals lead and chromium or silver, which may be too expensive.

Preparation of Reagent Solutions

The given weight of reagent is dissolved in water and made up to 100 mL. Use distilled water; otherwise, undesired insoluble chlorides will form in some experiments. Some of the reagents are used in more than one experiment; to allow you flexibility in choosing the gels you want students to do, directions are given for each solution used in each gel system.

1. **Copper Tree**
   1.0 *M* copper(II) chloride, $CuCl_2$, 13.5 g dihydrate
   1.0 *M* sodium chloride, NaCl, 5.9 g

2. **Lead Tree**
   1.0 *M* lead acetate, $Pb(OCOCH_3)_2$, 32.5 g, or 1.0 *M* lead nitrate, $Pb(NO_3)_3$, 33.1 g
   Thin pieces of zinc metal, Zn, about 0.5 × 1.0 cm

3. **Lead Iodide Ferns**
   1.0 *M* lead acetate, $Pb(OCOCH_3)_2$, 32.5 g
   2.0 *M* potassium iodide, KI, 33.2 g

4. **Lead Iodide**
   The gel: 1.0 *M* KI, 16.6 g
   1.0 *M* lead acetate, $Pb(OCOCH_3)_2$, 32.5 g

5. **Potassium Tartrate**
   The gel: 3.0 *M* tartaric acid, $HOOC(CHOH)_2COOH$, 45.3 g (Used in place of acetic acid.)
   Saturated potassium nitrate, KNO_3, 13 g, or KCl, 34.7 g

6. **Copper(II) Tartrate**
   3.0 *M* tartaric acid, $HOOC(CHOH)_2COOH$, 45.3 g (Used in place of acetic acid.)
   Saturated copper(II) sulfate, $CuSO_4$, 31.6 g (Copper(II) chloride does not work as well.)

7. **Copper Crystals**
   1.0 *M* copper(II) sulfate, $CuSO_4 \cdot 5H_2O$, 25 g
   1.0 *M* hydroxylamine hydrochloride, $NH_2OH \cdot HCl$, 7 g

8. **Copper(II) Reduction with Hydrazine Hydrochloride**
   1.0 *M* copper(II) sulfate, $CuSO_4 \cdot 5H_2O$, 25 g
   0.1 *M* hydrazine (as the hydrochloride, $N_2H_4 \cdot HCl$, 6.9 g

9. **Copper(II) Chromate/Copper(II) Hydroxide**
   (This gel sets up within 30 minutes.)
   1.0 *M* potassium chromate, K_2CrO_4, 20 g
   1.0 *M* copper(II) sulfate, $CuSO_4 \cdot 5H_2O$, 25 g

10. **Silver Chromate**
   (Note that this system uses different silicate and acetic acid concentrations.)
   1.0 *M* potassium chromate, K_2CrO_4, 20 g
   0.6 *M* CH_3COOH (3.5 mL glacial acetic acid)
   Sodium silicate (36.0 mL saturated solution + 84.0 mL water)
   1.0 *M* silver nitrate, $AgNO_3$, 17 g

11. **Silver Tree**
   Aluminum foil
   0.5 *M* silver nitrate, $AgNO_3$, 8.5 g

If your institution has the means to safely dispose of toxic mercury waste, the following two experiments are interesting ones to perform.

12. **Mercury(II) Oxide/Mercury(II) Chloride**
   Saturated mercury(II) chloride, $HgCl_2$, 6.9 g (*Hazard:* Mercury compounds are toxic.)

13. **Mercury Iodide (Four Systems)**
   A. 1.0 *M* potassium iodide, KI, 16.6 g
      1.0 *M* mercury(II) nitrate, $Hg(NO_3)_3 \cdot 2H_2O$, 34.3 g (*Hazard:* Mercury compounds are toxic.)
   B. 1.0 *M* potassium iodide, KI, 16.6 g
      1.0 *M* mercury(II) nitrate, $Hg(NO_3)_3 \cdot 2H_2O$, 34.3 g (*Hazard:* Mercury compounds are toxic.)
      Saturated sucrose solution, $C_{12}H_{22}O_{11}$
   C. 1.0 *M* acetic acid, $HC_2H_3O_2$, saturated with NaCl
      1.0 *M* potassium iodide, KI, 16.6 g
   D. Solid potassium iodide, KI, 2 to 3 g
      0.4 mercury(II) chloride, $HgCl_2$, 6.6 g (*Hazard:* Mercury compounds are toxic.)

Discussion

This long-term experiment merits exhibition outside the confines of the general chemistry laboratory. The tubes should be neatly labeled and displayed.

To seal the tubes, use high-quality corks, well softened. Because of the extended time over which this experiment takes place, it is necessary to watch for and correct inadvertent evaporation from bad corks.

As an independent research project it is of interest to study the effect of different molarities of the acid in forming the gel. Also, (see Postlaboratory Question 1) it is of interest to study the effect of temperature on this experiment.

Cleaning Up: Gels containing lead, chromium, and mercury must be disposed of as hazardous solid waste. Other gels can be disposed of in the solid waste container. In no case should gels be washed down the sink.

Alternatively, the gel may be dissolved away from the crystals with methanol. This will greatly decrease the volume of the heavy metal crystals, which can be removed by filtration. The methanol solution, after extensive dilution with water, can be poured down the sink.

Answers to Prelaboratory Questions

1. A 45% saturated solution contains 450 g of SiO_2, which is 7.49 moles:

   $$\frac{450\text{ g}}{60.09\text{ g/mol}} = 7.49\text{ mol}$$

   The number of moles of silicic acid is the same. The amount of silicic acid in the dilute solution is

   $$\frac{150}{1000} \times 7.49 = 1.12\text{ mol/L}$$

   This solution is diluted with an equal volume of acetic acid solution so that it is 0.56 molar in SiO_2 and thus contains 33.6 g of SiO_2 per liter. Thus the gel is about 3.4% SiO_2 or over 96% water.

2. Students should look up information on the following: Experiments 1, 7, 8, Cu; 2, Pb; 3 and 4, PbI_2; 5, $K_2(CHOH)_2O_4$; 6, $Cu(CHOH)_2O_4$; 9, $CuCrO_4$; 10, Ag_2CrO_4; 11, Ag; 12, HgO; 12, HgO and $HgCl_2$; 13, (A–D), HgI_2.

Answers to Postlaboratory Questions

1. An increase in temperature would increase the rate of diffusion of one solution into another and increase the rate of formation of the crystals and precipitates. But large crystals require slow growth rates, so the crystal size would decrease as the temperature rises.

2. It is likely that many of the gels would have the same appearance, because the formation of the crystals depends on the diffusion of one solution into another. Diffusion over the small spaces of this experiment will not be affected markedly by gravity. One would expect some slowing of crystal growth because convection would be slowed, but the whole purpose of the silica gel is to slow convection.

3. a. All odors detected by the nose are the result of the diffusion of a gas within a gas.
   b. Dust in the air, for example from a volcano, can diffuse around the world.
   c. If blue ink is carefully layered onto water, it will be seen to diffuse into the water over time. If temperature changes occur, the mixing of the layers will be greatly speeded up by convection in the solution.

EXPERIMENT

39 Synthesis and Analysis of Aspirin

Placement: Organic chemistry

Approximate time for completion: 2 hours

Approximate preparation time: 30 minutes

Materials: 24 of each, unless noted otherwise, assuming a class of 24 students working individually

Apparatus

Reaction tube or 10 × 100-mm culture tube with cork
Boiling stick
Steam bath, boiling water bath, or electrically heated sand bath
96-well test plate (flat-bottom preferably). See "Notes on the Procedure."

Reagents

Salicylic acid, 2 g
Acetic anhydride, 20 mL
Concentrated phosphoric acid, 4 mL
Distilled water, 25 mL
50% (v/v) aqueous ethanol (solvent for Part 2), 500 mL. Dissolve 250 mL of alcohol in 250 mL of distilled water.
Iron(III) chloride test solution. Dissolve 0.483 g $FeCl_3 \cdot 6H_2O$ in 100.0 mL of distilled water. The solution has a concentration of 1.0 mg Fe^{3+}/mL.

Discussion

This experiment involves the synthesis of a very widely used analgesic. It should not be ingested. The reaction, esterification, is a common organic reaction. It should be pointed out that vegetable oil and animal fats are esters.

This experiment serves as an excellent introduction to spectrophotometric analysis. The salicylic acid/Fe^{3+} complex follows Beer's law.

Notes on the Procedure

1. If available, 12-well strips are more convenient for the color comparison. You also can use 8-well strips, but this will reduce the number of comparison wells available and take away the blank wells at the end.

2. In the prelaboratory discussion, go through the analysis (Part 2) carefully. In particular, be sure students understand that the fainter the complex color, the better, since the complex forms as the result of incomplete reaction between salicylic acid and acetic anhydride.

Answers to Prelaboratory Questions

1. Anhydride means *without water*. Two molecules of acetic acid can be made to react to give acetic anhydride and one molecule of water. The acetic anhydride formed is without water.

2. Dissolve 2.703 g of iron(III) chloride hexahydrate in water and bring the volume up to 100 mL in a volumetric flask. The solution contains 0.027 g of the hexahydrate per milliliter. The hexahydrate contains 55.85/270.30% iron (atomic wt of Fe/formula wt of hexahydrate) so that each milliliter contains 0.00558 g or 5.58 mg/mL.

Answers to Postlaboratory Questions

1. The reaction of aspirin with water, even the moisture in the air, will hydrolyze the molecule into salicylic acid and acetic acid. The acetic acid can be detected by its vinegar odor.

2. The two compounds formed are acetic acid and salicylic acid.

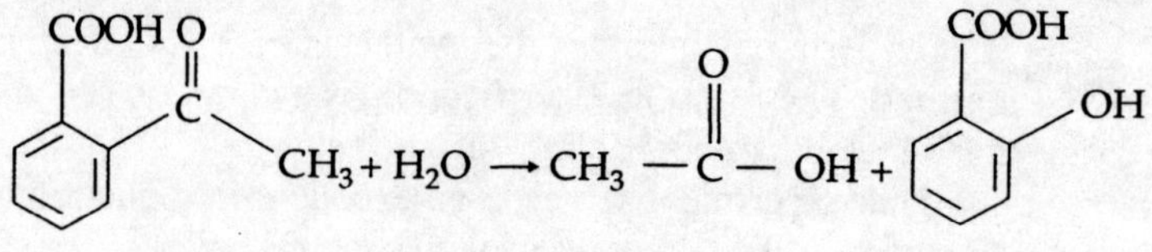

EXPERIMENT

40 Synthesis of Esters

Placement: Organic chemistry

Approximate time for completion: 2 hours

Approximate preparation time: 30 minutes

Materials: 24 of each, unless noted otherwise, assuming a class of 24 students working individually

Apparatus

10 × 100-mm reaction tubes or 10 × 75-mm culture tubes
Boiling chips
Wood boiling sticks
Sand bath and controller
Pasteur pipettes and rubber bulbs
1-mL graduated pipette and syringe adapter

Reagents

If all students make all combinations, you will need 10 to 15 mL of each alcohol and each liquid acid.

Methanol
1-Propanol
1-Butanol
iso-Amyl alcohol
Acetic acid
Propanoic acid
Salicylic acid, 3 g

Dowex 50X ion-exchange resin, about 1 g per class for each ester students make. The resin can be collected and reused, but it is most convenient to do so after the entire experiment has been completed. Expect about a 20% loss.
Sulfuric acid, concentrated, about 2 mL

Discussion

This is a favorite experiment with students because of the many familiar and pleasant odors produced. The procedure is simple, and students can make several esters in a short time.

Except for the butyl acetate and methyl salicylate combinations, we have left it to the student to calculate the amounts of reagents needed. Following Prelaboratory Question 1, 3.3 mmol each of the alcohol and organic acid is needed in each combination.

Placed fairly late in the year, the experiment makes a good review of electronegativity, polar bonds, and Lewis acid-base reactions, with the oxygen of the alcohol acting as the Lewis base and the carbonyl carbon acting as the Lewis acid.

Notes on the Procedure

1. The reaction mixture may bump because the Dowex resin covers the bottom of the reaction tube. It will help if the tube is heated very gently and inclined very steeply so that if the mixture does bump the material stays in the tube. It is important that students not point the open end of the tube at themselves or anyone else.

2. To reduce the risk of splashing, one of us (JGL) has had good success using Pasteur pipettes as reflux condensers. Wrap a few layers (depending on tube size) of Parafilm around the butt of the pipette, and then wedge it snugly into the mouth of the test tube after reagents and catalyst are in place. Slightly wider (that is, 13 × 100 mm) test tubes are a good choice if these condensers are to be used.

Answers to Prelaboratory Questions

1. $0.247 \text{ g}/74.12 \text{ g/mol} = 3.33 \times 10^{-3}$ mol.

2. $0.204 \text{ g}/61.05 \text{ g/mol} = 3.34 \times 10^{-3}$ mol.

3. $(3.33 \times 10^{-3} \text{ mol}) \times (60.1 \text{ g/mol}) = 0.200$ g;
$0.200 \text{ g}/0.804 \text{ g/mL} = 0.25$ mL.

Answers to Postlaboratory Questions

1. Aspirin is an ester because it contains the ester group, RCOOR'.

2. If 1 mole each of the acid and alcohol were allowed to react with each other, the concentrations of the ester, water, acid, and alcohol would all be 0.5 *M*.

EXPERIMENT

41 Oxidation-Reduction: Dyeing with Indigo, the Blue Jeans Dye

Placement: Physical properties, bonding, separation of mixtures

Approximate time for completion: 1.5 hours

Approximate preparation time: 20 minutes

Materials: 24 of each, unless noted otherwise, assuming a class of 24 students working individually

Apparatus

Electrically heated sand bath or hot plate
Thermometer (optional)
30-mL beaker
Three-prong clamp or small ring to support beaker
Glass rod or tongs
Ring stand
Metal spatula
Paper towels
Cloth for dyeing, 1 to 2 small swatches per student. (See note 1 of "Notes on the Procedure.")

Reagents

3 *M* sodium hydroxide, about 35 mL
Indigo, 1.5 g
Sodium hydrosulfite, 8 to 10 g (See note 4 of "Notes on the Procedure.")
Liquid or solid soap or detergent
3% hydrogen peroxide (optional). One small bottle of commercial antiseptic.
Household bleach, 5.25% NaOCl (optional). One small bottle of household bleach.

Discussion

Since most students throughout the world wear indigo-dyed clothes, this experiment is particularly relevant. There is a good explanation for the characteristic appearance of indigo-dyed denim, and the experiment is fast and relatively safe.

Notes on the Procedure

1. Multifiber cloth can be dyed quite well, and it is interesting to see how the different fibers react with indigo. Multifiber fabrics are available from Testfabrics, Inc., P.O. Drawer 0, 200 Blackford Avenue, Middlesex, NJ 08846; telephone: 908-469-6446; fax: 908-469-1147. They have a number of multifiber materials. We have had good success with Test Fabric Style 10 (6 fabrics, repeating every 1/3 in) and Style 43 (13 fibers, repeating every 1/3 in). The latter comes as a 4.5-in-wide roll, while the former is 48 in wide. If cotton or linen (cellulose) is used, it must be free of surface coatings in order to dye well. This means that permanent press fabrics will not work. A *very old* linen or cotton bed sheet torn into 2-in squares works very well.

2. The quantities of dye and sodium hydrosulfite are not critical. It is simply necessary to add enough of the hydrosulfite to reduce the indigo.

3. The dye bath can be warmed on the electrically heated sand bath, a hot plate, or least desirable, a Bunsen burner. In the latter case it should be supported over a ring and wire gauze with a large three-prong clamp or a small ring that surrounds the beaker. The temperature of the dye bath should be somewhere in the region of 80°C. An exact temperature is not critical.

4. Sodium hydrosulfite decomposes over time, so old reagent may not work in this experiment. Hydrosulfite of medium age will not be as active as the fresh material, and therefore, more than 0.25 g will be required. An excess does no harm. We have never experienced any safety difficulties from this reagent, although the label says it is flammable. Be particularly careful to purchase, stock, and use sodium hydrosulfite, $Na_2S_2O_4$. It is also called *sodium dithionite*. It is *not* sodium hydrogen sulfite, sodium bisulfite, or sodium thiosulfate.

Answer to Prelaboratory Question

1. Soap is added to wet and disperse the indigo throughout the water. Sodium hydrosulfite is used to reduce the indigo to the colorless form.

Data, Observations, and Conclusions

1. Immediately after the indigo is added to the water, it floats on the water.

2. After soap is added to the water and the mixture stirred, the indigo disperses in the water. It no longer floats on the surface.

3. When a drop of the suspension of indigo is placed on a piece of filter paper, the blue color remains at the center of the drop and the water spreads as a circle away from the center. A solution is homogeneous and would not exhibit this behavior.

4. After the sodium hydrosulfite is added to the suspension of the indigo, the color changes from the blue of the suspended dye to a clear yellow color.

5. The solution appears blue from above because the surface is in contact with air, which continuously oxidizes some of the leuco form to insoluble indigo, which floats on the surface. From the side the solution appears to be yellow.

6. When a drop of the leuco solution is placed on a piece of filter paper, it has little color at first, but as it air oxidizes, it turns uniformly blue. The fact that it turns uniformly blue indicates that the leuco form was in solution, in contrast to the suspended form of the dye.

7. Immediately after removing the cloth from the dye bath, it is almost white, with just a tinge of blue. While it was in the dye bath it could be seen that it was not colored.

8. The cloth gradually takes on a blue color over a period of time as the leuco-indigo is air oxidized to indigo. For reasons unknown, some of the fibers in multifiber cloth have a greenish color at first but eventually turn blue.

9. If indigo-dyed cloth is added to a hot solution of sodium hydrosulfite, we would expect the indigo on the surface of the cloth to be reduced and go into solution. In other words, the cloth would be bleached.

10. Hydrogen peroxide is an oxidizing agent, and indigo is the oxidized form of the leuco-indigo/indigo pair, so we would not expect peroxide to have an effect on the indigo.

11. Sodium hypochlorite is a reducing agent and would be expected to reduce the indigo on the surface of cloth, turning it into water-soluble leuco-indigo. In other words, the cloth would be bleached.

12. Each H, O, N. and C has a valence of 1, 2, 3, and 4, respectively; that is, each of these atoms has 1, 2, 3, or 4 bonds attached to it.

13. In the leuco form of indigo, the hydrogens that seem to have been added are attached to the two oxygen atoms.

14. Hydrogen bonds form between hydrogen atoms attached to nitrogen or oxygen and the unpaired electrons on other nitrogen or oxygen atoms. Fibers such as polypropylene, which is just a hydrocarbon, cannot form hydrogen bonds and so do not dye well (or at all) with indigo. Cellulose, with its many hydroxyl groups, can hydrogen bond to dyes

very easily. Wool and silk have even stronger centers where hydrogen bonds can form, so they dye exceptionally well. They also can form ion-dipole interactions. Nylon and Dacron, on the other hand, do not have a dense accumulation of polar oxygens and nitrogens; each of these fibers is mostly hydrocarbon in structure, so they pick up the dye less well than cotton but better than polypropylene.

Answers to Postlaboratory Questions

1. The reduction was carried out by fermentation, which is often a reductive process. It is possible to reduce indigo by adding it to an active ferment of sugar and yeast.

2. Many other dyes form covalent bonds to the fiber or dissolve inside the fiber and hence cannot be rubbed off.

EXPERIMENT

42 Synthesis of Slime

Placement: Hydrogen bonding, polymers, molecular geometry, organic chemistry

Approximate time for completion: 45 minutes, but students will enjoy playing with the product, so it will run over 1 hour

Approximate preparation time: 30 minutes

Materials: 24 of each, unless noted otherwise, assuming a class of 24 students working individually

Apparatus

Milligram balances (at least 1 for every 6 students, preferably 1 for every 4 students)
Paper cups, 3 oz, bathroom size)
Wood applicator sticks
10-mL graduated cylinder
1-mL graduated pipette (Mohr type or similar)
pH paper, hand-held pH meter, or universal indicator

Reagents

Polyvinyl alcohol solution, 4% w/v. Add 20 g of solid to 500 mL of distilled water. Heat to 70°C. Stir frequently, and maintain the temperature until the solution clears. Do not allow it to boil.

Discussion

The student text includes an extensive study of polymers in general, as well as of slime (polyvinyl alcohol) and silica gels (such as were used in Experiment 38, "Growing Crystals in Gels"). Unless they have models available, many students will have trouble visualizing the structures, so it may help to include some use of molecular models in your prelaboratory discussion. The idea of cross-linking (by borax) can be explained by analogy to a rope ladder, with the "rungs" being the borate units and joining the polyvinyl alcohol "ropes" via hydrogen bonding. For this reason, this experiment fits very nicely at the end of discussions of intermolecular forces and hydrogen bonding.

The synthesis of slime has, as stated in the "Introduction," much interesting chemistry involved in its formation. The 4% aqueous solution of poly(vinyl alcohol) is prepared in a large beaker equipped with a large magnetic stirring bar. The requisite amount of the poly alcohol (for example, Aldrich 36,314-6, poly(vinyl alcohol), 99+% hydrolyzed, MW 85,000–146,000) is poured in a steady stream into the rapidly stirred solution. It is allowed to stir for several hours until the powder has dissolved and a clear solution with the viscosity of cream results. Under no circumstances should water be added to the alcohol; an intractable goo results that will not dissolve in a reasonable period of time.

The monomer molecule, vinyl alcohol, does not exist, so vinyl acetate is polymerized and the resulting poly(vinyl acetate) is hydrolyzed to the corresponding alcohol. It is important to use the completely hydrolyzed material with an average molecular weight of more than 85,000.

The experiment is most conveniently carried out in the smallest paper cups available. Slime *is* soluble in water, but beakers containing slime are not rapidly washed clean.

Notes on the Procedure

1. The pH of a 0.2 *M* boric acid solution is 5.5. When 1, 2, and 3 mL of 0.2 *M* sodium hydroxide are added to 1.0 mL of this solution, the resulting pH values are 7, 10, and 12. Students can determine this information with pH meters (more solution will be needed) or indicating paper, or the information can simply be supplied to them. It is hoped that students will see the relationship between the pH of the solutions and the nature of the boric acid/borate equilibrium and then the nucleophilicity of the borate ion, which contributes so much to its ability to form hydrogen bonds.

2. Addition of 2 mL of boric acid solution to the PVA causes no change in viscosity.

3. Addition of 1 mL of sodium hydroxide solution to the mixture causes the viscosity to increase noticeably, but it is still a liquid.

4. Addition of another milliliter of sodium hydroxide causes a dramatic change, and true slime is produced. Thorough stirring is necessary; after a few minutes, the material can be removed from the cup.

5. When the concentration of hydroxide is increased even further by adding 1 mL of boric acid solution mixed with 3 mL of the 0.2 *M* sodium hydroxide solution to the PVA, a very firm gel is formed. When removed from the cup, it retains the shape of the cup but will, over a period of time, flow into a puddle about 6 mm thick.

6. When the slime from the first cup is heated to about 95°C, it (a) loses much of its viscosity and can easily be poured from the beaker. Because of poor heat conduction through the gel, about 5 minutes of heating will be needed. When the cool slime is rolled into a ball (b) and placed on a flat surface, it slowly flows into a puddle. When

struck sharply with the hand (c), the slime shatters. When picked up and allowed to fall by its own weight, it will stretch (d) into a long, thick thread that can be more than 30 cm long. When rolled into a cylinder and pulled sharply (e), it will break, and it will bounce (f), although not as well as a rubber ball or silly putty.

7. The slime does dissolve in water, albeit slowly. The beaker from the previous experiment should be washed out at this time.

8. When sodium chloride is added to the mixture, the gelatin process does not occur nearly so well, and a reduced viscosity mixture results. This is presumably because the sodium ions and the chloride ions are competing with the borate ion for poly(vinyl alcohol) and water. The ions are said to be solvated by the water and the alcohol.

Sodium Silicate Experiments

9. Sodium silicate solution is made by dissolving sand (silica, SiO_2) in sodium hydroxide. The material sold by Aldrich is 14% sodium hydroxide and 27% SiO_2. As such, it is quite corrosive and should be handled accordingly.

   Silicon has an electronegativity, 1.8, similar to that of boron, 2.0, and is located just below carbon in the periodic table, while boron is located just to the left of carbon. Silicic acid, $Si(OH)_4$, is therefore analogous to boric acid, $B(OH)_3$, but unlike boric acid is unstable and readily polymerizes to silica gel. The silicate ion is tetrahedral. Removal of some of the hydrogens in strong base gives sodium silicate, also known as *water glass*. This material is capable of forming very strong hydrogen bonds to poly(vinyl alcohol). The gel formed from 2 mL of sodium silicate and 10 mL of PVA is quite rigid and becomes more rigid over a 24-hour period. The gel formed from equal parts of sodium silicate solution and PVA is rigid enough to be cast into small molds, such as candy molds.

Answers to Prelaboratory Questions

1. Student answers will vary with regard to pH because, as the text points out, the hydrogen bonding is possible with either boric acid or borate ion. The structures and discussion focus on borate first, and since addition of NaOH favors borate, most students should predict that the gel should get stronger. They may or may not reason that a stronger gel is less slime-like. The text is very clear on the fact that heating weakens the hydrogen bonds.

2. Since pK_a for water is 14 and for boric acid is 9.2, boric acid is the stronger acid. This means that $B(OH)_4^-$, its conjugate, is the weaker base. The conjugate acid-base pairs are $H_3BO_3/B(OH)_4^-$ and H_2O/H_3O^+.

Answer to Postlaboratory Question

1. Student answers.